INTELLIGENT MANUFACTURING

Reviving U.S. Manufacturing Including
Lessons Learned from Delphi Packard
Electric and General Motors

INTELLIGENT MANUFACTURING

Reviving U.S. Manufacturing Including Lessons Learned from Delphi Packard Electric and General Motors

R. Bick Lesser

CRC Press
Taylor & Francis Group
Boca Raton London New York

CRC Press is an imprint of the
Taylor & Francis Group, an **informa** business

A PRODUCTIVITY PRESS BOOK

DiPietro Library
Franklin Pierce University
Rindge, NH 03461

The views and opinions expressed herein are solely those of the author and do not necessarily reflect the views of Taylor & Francis LLC, CRC Press, Productivity Press, or any of its affiliates or employees.

CRC Press
Taylor & Francis Group
6000 Broken Sound Parkway NW, Suite 300
Boca Raton, FL 33487-2742

© 2014 by Taylor & Francis Group, LLC
CRC Press is an imprint of Taylor & Francis Group, an Informa business

No claim to original U.S. Government works

Printed on acid-free paper
Version Date: 20130729

International Standard Book Number-13: 978-1-4665-6404-6 (Paperback)

This book contains information obtained from authentic and highly regarded sources. Reasonable efforts have been made to publish reliable data and information, but the author and publisher cannot assume responsibility for the validity of all materials or the consequences of their use. The authors and publishers have attempted to trace the copyright holders of all material reproduced in this publication and apologize to copyright holders if permission to publish in this form has not been obtained. If any copyright material has not been acknowledged please write and let us know so we may rectify in any future reprint.

Except as permitted under U.S. Copyright Law, no part of this book may be reprinted, reproduced, transmitted, or utilized in any form by any electronic, mechanical, or other means, now known or hereafter invented, including photocopying, microfilming, and recording, or in any information storage or retrieval system, without written permission from the publishers.

For permission to photocopy or use material electronically from this work, please access www.copyright.com (http://www.copyright.com/) or contact the Copyright Clearance Center, Inc. (CCC), 222 Rosewood Drive, Danvers, MA 01923, 978-750-8400. CCC is a not-for-profit organization that provides licenses and registration for a variety of users. For organizations that have been granted a photocopy license by the CCC, a separate system of payment has been arranged.

Trademark Notice: Product or corporate names may be trademarks or registered trademarks, and are used only for identification and explanation without intent to infringe.

Library of Congress Cataloging-in-Publication Data

Lesser, R. Bick.
  Intelligent manufacturing : reviving U.S. manufacturing including lessons learned from Delphi Packard Electric and General Motors / R. Bick Lesser.
       p. cm.
  Includes bibliographical references and index.
  ISBN 978-1-4665-6404-6
  1. Manufacturing industries--United States--Management. 2. General Motors Corporation--Management--Case studies. 3. Delphi Corporation--Management--Case studies. I. Title.

HD9725.L47 2013
658--dc23   2012023772

Visit the Taylor & Francis Web site at
http://www.taylorandfrancis.com

and the CRC Press Web site at
http://www.crcpress.com

Contents

Introduction ix

1 **What Would You Expect with a Parent Like GM?**....................... 1
   Interview Process 2
   Packard Electric Is Born 3
   How Could GM (Packard's Parents) Make So Many Dumb Decisions? 7
   So, How Are Things Working Out for GM Today? 12
   How Does GM Compete with Toyota and the Vaunted
   Toyota Production System? 15

2 **We Have to Do Something, Even if It's Wrong** 19
   Packard Does a Really Smart Thing Next 21
   Where Did the Packard Training Program Originate? 23
   Good News and Bad News about Packard's Management 27
   A Move to the Warm South 29
   What We Need Are Suppliers 30
   Mexico, Here We Come 33
   Black Friday 33
   Something Had to Change for GM 36
   Packard Goes International in a Big Way 37
   We Are Not Ready for This 38
   We Are in Real Trouble Here 40
   Industrial Engineering (IE) Training Program 43
      The Industrial Engineering Training Program 44

3 **How Does a Company with So Many Smart People Do So
   Many Dumb Things?** 49
   Automation Can Save Us 51
   That Worked So Well, Let's Try Something a Little Tougher 53
   That Lead Prep Startup Surely Didn't Work Out Like We Had Planned 56

Integrated Production System (IPS) to the Rescue (or Not)......................59
If IPS Didn't Get It Done, IPS II Surely Would 67
Remote Lead Prep: Yeah, That's the Answer!...70
How about Just Getting Rid of Our Productivity Control System............74
So Just How Are IDCs Going to Help Us with the 3% Give Back? 77
How about Let's Just Get into a New Business 79
Let's Sum Things Up..80

4 Let's Get Lean ... Not 83
Just What Is the TPS?...85
Toyota Production System Rules 98
Packard Develops the PPS...100
    What We Need Is Vista 100
    Benchmarking: To Do or Not to Do 102
    Kitting Looks Like a Cool Thing to Do..105
    Let's Do the Team Concept Like Toyota..109
    How about a Nice Quality Circle?...110
    Did QS9000 Implementation Really Help? 112
    Say Zero Defects and Mean It..114
The Good News Is That We Saved $25 Million in Inventory, but 115
Just-in-Time Manufacturing?...118
U-Cells Gain Great Prominence 120
Somebody Finally Got One-Piece Flow Right ... Really?.......................122
Wrap-Up...128

5 Proper Manufacturing Organization Is Critical (*or* You Will Get the Performance You Motivate).....................131
Plant Priority Meetings...132
Let's Just Change Reporting Lines to Get Better Support......................133
Warren Got Some Things Right 139
Understanding Human Nature...139
So We Transferred Our Manufacturing Successes to Mexico, Right?....143
What's the Fix? 147
Manufacturing Supervision Is Tough..150
    Providing Proper Manufacturing Support...150
        Why Worry about Improved Productivity, Mexican Labor Is Cheap (and Philippine)..151
    Wiring Harness Departmental Organization Basics 152
        Level 1 Operators 153
        Level 2 Operators 153

Contents ■ vii

    Level 3 Operators 154
    Level 4 Operators 155
    Foreman's Responsibilities and Support Required 156
    Job Responsibilities 156
    Time Breakdown 157
    Assistant Foremen/Group Leaders 158
    Additional Advice 158

6 When You Measure Performance, Performance Improves 161
  From Routings to Direct Labor Bibles (DLBs) 162
  Production Efficiency 164
  Process and Operator Efficiencies 166
  Wouldn't One Efficiency Number Be Sufficient? 168
  The Warren Bottom Line 169
  So, What Did We Do in Mexico? 170
  I Finally Got a Chance to Do It Right in South Korea 171
  How Much Could the Mexican Operations Have Saved? 177
  When You Measure Performance 182
  Packard Productivity Nose Dives 184
    Are We Really Controlling Productivity in Our Plants? 184
    Direct Labor Bibles (DLBs) 185
    Efficiencies from a Budget Routing 186
    Plant Productivity Information Available Based on Labor
    Estimates (Budget DLBs) Only 186
    Plant Productivity Information Available with an Automated
    Productivity System Based on Production DLBs 188
    Terms and Definitions Used with the Computerized
    Productivity System 190
    Simple DLB Example 192
        Metal Cutting 193
        Component Subassembly 195
        Final Assembly of a Widget 195

**7 Preplanning: The Perfect Tool to Accomplish Toyota's
   Rule #1 .... 199**
  What Is the Common Industry Practice? 200
  But, Is It Worth It? 202
  Packard Evolves, but Not for the Better 205
  The Critical Four M Relationship 207
  Wiring Harness Preplanning 208

8   The Computer Is a Moron 219
    A Rare Opportunity 222
    Controlling Absenteeism, Turnover, and Overtime Is a Must 226
    Material Availability Is Critical 231
    Summary 235

9   How to Drive Down Total Process Cycle Time (TPCT) without Wasting a Lot of Time (and Money) 237
    Critical Value Streams 239
    Packard "Attacks" the TPCT for Engineering Change Implementation 240
    Key Activity Control Is Born 242
    Others Benefit from the System 248
    This System/Concept Will Work Anywhere 255
    The Bottom Line 258

10   What Size Should the Cycle Quantity (Lot Size) Be? 259
    Portugal Will Show the Way 263
        Cost Bucket One 264
        Cost Bucket Two 266
    Some People Don't Want to Be Confused with Facts 267
    Free Rein in South Korea 270
    Determining Optimum Cut Quantity (or Optimum Process Quantity) 272
    Summary 279
        Optimum Cut Quantity Subsystem 280
        Optimum Cutter Loading and Sequencing Subsystem (OCLASS) 286

11   Wrap-Up: How about "Intelligent Manufacturing" for Real Change in Which We Can Believe 291

Index 301

About the Author 309

Introduction

Why Is U.S. Manufacturing in Such Bad Shape?

During most of the past century the United States has been revered as the world's great industrial giant, and it hasn't been close. Not only has the United States been the country in which the most new products have been invented, but also it has been the country in which a large part of the world's manufacturing capacity existed. The United States is still the world's largest manufacturer, but China has dramatically closed the gap, and, undoubtedly, will continue to do so until it overtakes this country, unless something drastic is done.

It becomes harder and harder to find U.S. manufactured goods, especially in consumer electronics and other consumer items, even when you look for them because you want to be patriotic and buy American. Just go into any big box store and try to purchase all of the items you need to run your household with products manufactured in this country. I remember that 30 years ago it wasn't a problem; today, it is practically impossible.

It's not because these retailers are anti-American, quite the contrary. They have generally shown themselves to be very responsible and generous in all of the communities in which they serve the public. The problem is that, in so many product lines, U.S. manufactured goods are just not available or, if available, are not competitive from a cost and/or quality standpoint.

Alexander Hamilton, who became the first Treasury Secretary and who had served as an aide to General George Washington during the British blockades, resolved that the United States would never again have its independence, freedom, and survival dependent on imported goods, such as French muskets and ships.

In 1791, after becoming Treasury Secretary, Hamilton wrote, "Not only the wealth, but the independence and security of a country, appear to be materially connected with the prosperity of manufactures. Every nation … ought to endeavor to possess within itself all the essentials of a national supply. These comprise the means of subsistence, habitation, clothing, and defense."

Under his leadership, the United States created a national free trade zone and implemented the tariff system for all imported goods. These duties gave the national government all of the revenue necessary to build the country's infrastructure, provide for the national defense, and create a manufacturing base that would become the envy of the world. By 1890, the United States had overtaken Great Britain as the world's leading manufacturer. A few years later, on the brink of World War I, the United States manufactured more than the combined outputs of Britain, France, and Germany.

Those of you familiar with World War I and World War II know that the manufacturing might of the United States was likely the deciding factor in the Allied victories. The world would be a very different place had this country not been in a position to come to the aid of its friends with the weapons of war in all varieties.

Pat Buchanan wrote an article entitled "Death of Manufacturing," which was published in the August 11, 2003 issue of *The American Conservative*. In this article, he laments the state of U.S. manufacturing (which has gotten nothing but worse since then) because he understands the importance of manufacturing to a country's national interests. He said, "Manufacturing is the key to national power. Not only does it pay more than service industries, but the rates of productivity growth are higher and the potential of new industries arising is far greater. From radio came television, VCRs, and flat-panel screens. From adding machines came calculators and computers. From the electric typewriter came the word processor. Research and development follow manufacturing."

Undoubtedly this is true, so the data provided by Pat Choate, author of *Agents of Influence* (Simon & Schuster, 1990), should be of great concern to each of us. He gives the following levels of U.S. dependency on foreign suppliers for critical goods:

Medicines and pharmaceuticals: 72%
Metal working machinery: 51%
Engines and power equipment: 56%
Computer equipment: 70%
Communications equipment: 67%
Semiconductors and electronics: 64%

And, these numbers were from 2003. Undoubtedly, they are worse today.

Buchanan also points out in "Death of Manufacturing" that one third of our labor force in 1950 was engaged in manufacturing. By 2003, that number had dropped to 12.5%. Now it is less than 10%. This is not comparable to the fact that only 2% of today's workers are engaged in agriculture as compared to over 90% 150 years ago. Our agricultural practices are the best in the world, and it only takes a very small percentage of our workforce to feed this entire country as well as much of the rest of the world. In the case of manufacturing, we have been losing thousands of good jobs every month for decades, not because they are no longer necessary, but because they have gone to other countries that are more competitive, and then these products are sold back to us, further deepening the U.S. trade imbalance.

Speaking of trade imbalances, there was a time that the United States ran a trade surplus and was also energy independent. Now this country is the world's largest debtor nation and the trade imbalances are growing exponentially. A few short years ago, this country was running trade deficits of about $30 billion per month, and that sounded to me like a lot of money then.

Until the recession hit U.S. consumers hard, America was running monthly trade deficits twice that. Data shows that our 2008 trade deficit (of something over $700 billion) was about equal to the sum of all of the trade deficits of all other deficit nations combined (Britain had the second biggest trade deficit, which was about a third of this country's). China had the largest trade surplus (something over $300 billion), followed by Germany and Japan. The situation has not gotten any better, judging by 2011 data. The U.S. deficit was $785 billion followed by Great Britain with a $163 billion deficit. Saudi Arabia took over the top spot as the world's leading creditor nation with a trade surplus of $253 billion (thanks to our purchases of its oil). Germany came in second with a $220 billion surplus. Russia was third with a surplus of $199 billion and China slipped to fourth with a $155 billion surplus.

The good news, as Pat Buchanan points out, is that we do have a trade surplus with China on a few things—most notably soybeans, corn, wheat, animal feeds, meat, cotton, metal ores, scrap, hides and skins, pulp and waste paper, cigarettes, gold, coal, natural gas, mineral fuels, rice, tobacco, fertilizers, and glass.

Does anyone truly believe that running a trade deficit of more than a half trillion dollars per year (created largely by America borrowing money from China to buy goods made in China) is a sustainable scenario? I'm no economist, but it doesn't make sense to me, and I'll bet it doesn't to you either. Just what is going to change the situation? A very appropriate quote, which has been attributed to Einstein, is as follows: "Definition of insanity: Doing the same things you have always done, the same way you have always done them, and expecting that the results are going to be different."

Competitive Position

I mentioned earlier that we continue to hemorrhage manufacturing jobs (in 2003, Buchanan says, 80,000 per month, and it has gotten continually worse) to countries whose competitive position is stronger than ours. This competitive position is comprised of several different factors, some of which are fairly earned, but many of which are not.

Buchanan's position is that we have lost our once superior place in the world of manufacturing through unfair trade practices, namely, free trade. In commenting about the unfairness of the current trade policy, especially as it pertains to China, Buchanan writes, "Hamilton, Clay, Lincoln, and Teddy Roosevelt would recognize China's policy for what it is and counter it. But, this generation of free traders does not have a clue as to what is going on, or does not care. Either way, the consequences will be the same: deindustrialization of America, declining of the dollar, a deepening dependency on foreign countries for the necessities of our national life, diminished sovereignty, and eventual loss of our independence.
If you disbelieve this, look at the once sovereign and independent nations of Europe."

I disagree with Buchanan on some issues, but it is pretty hard to argue with his point that "free trade does not work if you are the only one practicing it fairly." At least it doesn't work for you in the long run, even though consumers may benefit in the short term. I think most people, me included, believe that trade is good and isolationism, especially for the United States, is not possible or practical to practice in today's world. However, we are past the time where we can be benevolent and give every other country advantages in trade policy.

Having worked in Mexico for a dozen years, in Portugal for three, and in South Korea for four, as well as having spent quite a bit of time in Europe, South America, and Southeast Asia, I can say that our goods do not have the same opportunities in their markets that theirs do in ours. For Buchanan, this is virtually all due to unfair trade practices with which the United States has to contend. This is certainly part of the problem, but, as I see it, based on my international experience, there are four primary factors why our trade imbalance in manufactured goods is so out of control:

1. Unfair trade practices
2. Taxes
3. National preferences and attitudes
4. Quality, desirability, and cost of products

When these four factors are combined with the so-called U.S. energy policy (one which refuses to exploit our own energy sources in favor of expensive imports), we end up with the world's largest trade deficit.

I don't claim to be a trade expert, but some of the problems we are experiencing with our trade imbalance seem pretty straightforward. Having lived and worked and spent a good deal of time outside the United States, I think I have some additional insight that might be helpful.

Unfair Trade Practices

In his article, Buchanan points out that free trade is not fair trade, and we have certainly gotten the short end of the stick in our dealings with just about every other country on Earth. We have about as much chance of getting a fair ruling in the World Trade Organization (WTO), where Europe outvotes us 15 to 1, as we do of getting a favorable ruling in the United Nations, of which I believe the United States should no longer belong.

I believe our politicians sometimes tend to make things much too complicated. **Free trade** does not work, but **fair trade** would certainly help level the playing field.

The United States should implement a trade policy that I call "The Platinum Rule." It's not quite the Golden Rule, which would say that we should treat our trading partners as we would wish to be treated. It would be nice if that would work, but it won't. I totally agree with the Golden Rule when it comes to treating our family, friends, and neighbors, but it is very naïve to think this should be the basis of our trade policy with other nations.

The Platinum Rule would state that we should treat our trading partners as they deserve to be treated, and I believe this could be done in an objective and understandable manner. It probably wouldn't be received too favorable by some of our trading partners, but, then, most of them don't think too highly of us as it is.

I believe the right answer is to develop a Trade Factor for each of our trading partners that would take into account several factors pertinent to achieving fair trade. This factor would include such things as the openness of their markets, the level of tariffs they place on the goods we export to them, whether or not they float their currency, their valiance in the protection of intellectual property rights, their level of government subsidies, whether or not they practice dumping, their human rights record (child labor laws, workers rights, etc.), their record regarding industrial espionage, their trade relations with rogue nations, their level of pollution, and maybe another factor or two.

In other words, a country should have to earn a good trading relationship with the United States based on its performance. One of the themes that I will be using throughout this book is that people (or countries, for that matter) will do what they are motivated to do, or, maybe better said: **You will get the performance that you motivate.** If we want to motivate fair trade practices on the part of our trading partners, our trade policy better motivate that behavior.

I think each of us should be very concerned about some of the trading relationships we have made over the years, most especially the current relationship we have with China. China falls far short of the ideal on just about every factor listed above (which I think should be of grave concern to each of us). However, they are given favored nation status as a trading partner, when, in my opinion, the United States should be placing very high duties on all of the products imported from them. Can someone please explain this to me so that it makes sense? Have we considered the long-term consequences of the current policy and what the ultimate outcome is likely to be if we don't make a change regarding our relationship with China? Have we considered the probable long-term impact on our children and grandchildren?

Taxes

Regarding tax policy and its impact on trade, the United States is its own worst enemy, according to the Organization for Economic Co-operation and Development's 2011 data. Only Japan has a higher combined corporate tax rate. Does it make any sense at all that we have the second highest corporate tax rate of all industrialized nations, second only to Japan? It's good to be No. 1 in a lot of things (like the Olympics or the standard of living), but the corporate tax rate is not one of them.

I know it makes for good sound bites when the politicians chastise big business for not paying its fair share. I guess it is supposed to make the little guy feel better knowing the big guys are getting hammered. There is only one little problem: Corporations do not pay taxes. To corporations, all taxes (embedded, income, property, etc.) are a cost of doing business. All companies must make a return on investment that is acceptable to the owners; that is why all taxes get factored into pricing. **Consumers are the ones who pay all of a corporation's taxes.** If a corporation cannot pass on all taxes to the consumer in the form of higher prices and still earn a reasonable return for its owners, that corporation goes out of business.

It also is true that a company that pays higher taxes, embedded or otherwise, is in a less competitive position than a company that pays lower taxes, and it will have a decided disadvantage in international trade. If we really want to grow our economy and protect American manufacturing jobs, shouldn't our tax rate be zero for corporations? This would lead to lower prices for the American consumer and increased industrial exports (see below).

While we are on the subject of taxes, here would be a good place to inject that the abolishment of the income tax in favor of the Fair Tax would be a great way to eliminate all corporate income taxes as well as our own. Among the problems with the current system are that we are disincentivizing productivity, savings, and investment (three of the most critical things needed to support a growing and vibrant economy); not taxing a large part of the country's economic activity; creating an unbelievably expensive and powerful bureaucracy; giving the government a tool to control our behavior; requiring Americans and companies to spend billions on nonvalue-added recordkeeping and tax preparation; allowing a large percentage of Americans to escape any financial accountability for being citizens of this great country; putting our companies at a significant disadvantage to other companies around the world; and these are just a few of our current tax system problems.

National Preferences

The issue of national preferences becomes a little more difficult hurdle to overcome. Most Americans are patriotic and love their country and want America to be successful, not only for their own future, but for that of all Americans and their descendents. Most of us also believe in our system of capitalism, which has produced one of the highest levels of prosperity and freedom in the history of the world.

Our belief and confidence in capitalism dictate that we reward (patronize) businesses that do a great job of providing high-quality goods and services to the marketplace at reasonable prices. For most Americans, the decisions on which goods and services to purchase are based on these criteria, and the country of origin is no better than a tie breaker, if it is considered at all. Most Americans, I believe, do not feel unpatriotic in the least if they choose a product of European or Asian origin over a similar product of American origin if they "perceive" the foreign product to be "superior."

This is certainly not the case in many countries with whom we trade. There is no stigma attached to buying an imported product in America, but this is certainly not universal, especially in Asia, based on my personal experiences and observations. Some of this is due to covert (and sometimes overt) campaigns on the part of the governments, some of it is due to cultural traditions and norms, and some of it is due to peer pressure. Whatever the reasons, attitudes are not the same around the world. In some markets, U.S. products have to be demonstrably superior to even get consideration from a sizeable portion of the populace.

I don't have all the answers, but when it is evident that prejudices do exist that do not give our products a fair shot in the marketplace, we must work with those governments to reverse the unjustifiable biases. Probably the best way to do this is to ensure that our goods are demonstrably superior to others available in their marketplaces. This book may be able to help in this regard.

U.S. Energy Policy

It is a little difficult to talk about U.S. energy policy because it is not apparent that we have one, except that we don't seem to have a problem importing about 60% of our petroleum needs. You would think with all of this feigned concern about our trade imbalance, the high cost of energy, and the obvious impact on national security, it would be a paramount issue.

Oh, I know that there has been talk of the need to become energy independent, but that was before the 2008 and 2012 elections. Now, it seems like business as usual while we wait for wind, solar, and thermal power to become viable. I hope no one has any hopes of "green" energy sources helping us gain our energy independence any time soon.

Of course, in addition to reasonable conservation, the right answer is to drill in the Alaska National Wildlife Refuge (ANWR) and off the coasts, where there are large, proven reserves that are now off limits; to develop the gigantic oil shale deposits in the Western states; to continue to develop clean coal technology; to expand the use of natural gas; and to greatly expand the use of nuclear energy. The problem is that those in Washington don't like fossil fuels, and they are not too keen on nuclear power, either.

No rational person would argue against reasonable conservation measures and the need to develop alternative, renewable energy sources. But, putting our fate in the hands of world leaders, many of whom are not U.S. allies, who happen to be sitting on most of the world's oil reserves that are being developed, while we wait a generation or two to find alternatives to fill our energy needs, seems to be more than foolish.

I am confident we can find viable alternative energy sources over time, but history has clearly shown that this can be best accomplished through the private sector. We will get the behavior we motivate. If the government wants to be helpful in this process, the best thing it could do would be to offer large cash prizes (nontaxable, at that) to any entities discovering, developing, or inventing viable renewable energy sources (in addition to abolishing the Department of Energy).

And, this brings me to the last major problem on my list that is keeping us from closing the trade balance gap, which is resulting in the deindustrialization of America: desirability, quality, and cost of U.S. manufactured goods.

Desirability, Quality, and Cost of U.S. Manufactured Goods

This is the problem to which most of this book is devoted. I won't speak a great deal about desirability, except to say that the desirability of American-made goods is greatly affected by both quality and cost. The spirit of invention is still strong in America, and if we can solve the other problems listed in this introduction, as well as ensure good quality at a fair price, desirability won't be an issue (and, believe it or not, there are some foreigners who have a bias in favor of American-made goods).

The objective then is to ensure that U.S.-manufactured goods are "world class" in both cost (which determines price, if a company is to stay in business) and quality (I also can throw delivery into the mix, which, in large part, is determined by the success of the other two factors).

I have had advantages, which most American business people have not had, of being able to observe and to experience firsthand manufacturing operations around the world. Having lived and worked for years in manufacturing operations in the United States, Mexico, Portugal, South Korea, and the Philippines, and having spent considerable time observing and working with manufacturing operations in Germany, England, Ireland, Honduras, and Japan, I have a very good insight into the competencies of manufacturing operations of other nations, as compared to the United States.

Good News, Bad News

My observation is that our manufacturing competitiveness is a "good news, bad news" scenario.

Rush Limbaugh often says that America is the greatest country on the face of the Earth because of her people. I think many people misunderstand him and believe he is saying that Americans, as a collective, are superior beings to citizens of other countries. Being an American, it would be nice to think so; however, what I discovered, after spending so much time outside of the United States, is that there is no discernable difference in the average capabilities of people among the various nations with which I have had experience, especially among manufacturing workers. This is certainly the case among developed nations, but it is also true within less developed nations.

In other words, a Mexican is just as capable of running a molding press or a wire cutter or in stringing and plugging wire as is a German, who is just as capable as an Irishman, who is just as capable as a Japanese, who is just as capable as an American, who is just as capable as a Honduran, who is just as capable as a South Korean, etc., etc., etc.

I also have experienced that the same thing is true when dealing with engineers and other salary workers. In each of the countries in which I have worked, I have encountered exceptional employees, worthless employees, and everyone in between. Country of origin just doesn't seem to be a factor in the overall qualifications of employees, either hourly or salary. And, this would certainly make sense when considering the United States, because all Americans, with the exception of Native Americans, have ancestry from somewhere else.

So, the bad news is that we Americans are innately no better (although, no worse, either) than workers from other nations. However, there is good news.

Although we, as individuals, are not superior to individuals from other countries, the highly competitive free market system and advanced infrastructure within our country enable our manufacturing operations to be more productive than those of other countries. This is not only my contention based on extensive observation, evaluations, and study, but there are also scientific studies, including one by the McKinsey Global Initiative, showing this to be the case.

Based on my Packard experience during the 1970s through the 1990s, comparing apples to apples, if Packard's U.S. operations were assigned a productivity score of 100, the score of Packard's German operations would be about 80; the Portuguese, Irish, English, and South Koreans score about 70; and the Mexican score about 50. And, none of these differences in scores is due to differences in capabilities on the part of the employees. Each of the differences can be explained based on management and manufacturing systems in place, controls used, cultural norms, and levels of absenteeism and turnover. I will go into an explanation of these factors during the course of this book.

These observations on the differences in productivity levels between countries are in harmony with the findings of a study done in 1989 by the McKinsey Global Institute. They measured the productivity levels (in dollar value of goods produced in relation to the hours of labor required to produce them) of five of the world's industrial powers, i.e., the United States, Germany, France, Britain, and Japan. They discovered that the United States had the highest productivity level, to which they assigned a value of 100%. Germany came in at 80%, as did France. Japan came in at 76% (probably a surprise to most), and Britain came in at 61%. (The good news is that workers in healthcare and real estate, and, most pointedly, government and education, were not included in this study, or our numbers might not have looked so good.)

Most people reading this book will be surprised at the number for Japan, although it is in line with observations I made when spending time in that country. Japan has focused attention for decades on automobiles and consumer electronics and leads the United States in these areas, but the rest of the economy lags behind. As the McKinsey Institute determined, Japanese factory workers only produce 80% as much as Americans on an hourly basis and productivity in general merchandise retailing is only 44% of that of American workers.

After conducting and analyzing the study, William Lewis, director of the McKinsey Global Institute, said: "There's a lot of talk about how the United States should adopt a model from another country if economic operations in those countries were proving to be more productive. But, we found no evidence of that, so it is not obvious the United States should look to copy a model from somewhere else. It's more a case of making this model work better." (This statement was made well *after* the introduction of the Toyota Production System, Lean Manufacturing, and other similar concepts.)

And, this is the real focus of this book.

I spent over 30 years with a company that didn't have the wisdom to take what it was doing well and focus on making steady improvements, while borrowing sparingly from others based on careful analysis and trial. But, instead it kept looking for the silver bullet that was going to magically transform it into a "world class" company, all the while digressing in or discarding things it had been doing well all along. The end result being that Packard was a much stronger manufacturing company 30 years ago than it is now, despite millions of dollars being spent on consultants and various "Lean" initiatives.

Delphi Packard recently emerged from almost four years in Chapter 11 bankruptcy, along with the rest of the Delphi Corporation (formerly Delphi Automotive Corporation). The fact that Delphi Packard has digressed as a manufacturing company is not the primary reason that Delphi Corporation had to declare bankruptcy, but it certainly didn't help either. In my opinion, the bigger culprits are the unions and their entrenchment ideology, along with decades of ineffective management by its former parent company, General Motors, and to a lesser degree its own management.

However, there is a lot that can be learned from the experiences of Packard. One of my favorite sayings is: "It is a smart person that learns from experience, but it is a truly wise person who learns from the experiences of others." My hope is that by reading this book, which shares several proven new concepts, as well as lessons learned from Packard and GM, I can help each reader become a much wiser business person who will avoid making similar mistakes, while capitalizing on some of the things Packard did well.

Being a fairly innovative and industrious person with wide experience, I have taken the opportunity over the years to develop some concepts and tools that I am sharing in this book; concepts and tools that, if used, will help lead to "real" performance improvements in a manufacturing operation (and many of them also are applicable to nonmanufacturing businesses, as well).

So, What Is the Bottom Line?

America is part of a global economy, and that is not going to change anytime soon. This has provided us, and will continue to provide us, with many benefits, both as a consumer and as a manufacturing nation; however, it also presents some very difficult challenges.

The bad news is that as individuals we are not exceptional, but the good news is that as a nation we are because of the systems put in place by our Founding Fathers and perpetuated by our ancestors. Excluding the high cost of labor and the impact of unions (agreeably, two very interrelated and important factors), we are undoubtedly the most competitive nation on Earth (sometimes in spite of ourselves). However, we do have unions. (Hopefully, in the future, both unions, as well as management, will be focused more on the long-term viability of our companies and our country than on immediate gratification and personal gain.) And, we do have wages significantly higher than most of our competitors, something often strongly influenced by government policy, both in the United States and around the world. Eventually, if we are to have continued success as a manufacturing nation, we must allow the free market to dictate the value of labor, not have it artificially dictated by governments, unions, and special interest groups.

All of this means that we Americans must work smarter (not necessarily harder) than our competition (something Packard management often challenged, but rarely practiced). This book should help in that endeavor. If an individual or a company only incorporates one of the many concepts available in this book, the cost of this book, and the time and effort required to read and ponder it, will be well worth it.

It also means that we need our government to finally step up to its responsibilities to help resolve the other four problems stated earlier, which, together with the desirability, cost, and quality of American products, are responsible for our gigantic trade imbalances, imbalances that are certainly unsustainable. If we want the government to act, it must hear from us and hear from us in a big way. With politicians, it is the squeaky wheel that gets the grease.

About Rush Limbaugh's Statement

Now is a good time to clarify Limbaugh's statement, a clarification with which I believe he would agree. America is one of the greatest nations in the history of the world, not because of the American people per se, but rather because of the inspired system of government that was put in place by our Founding Fathers. The guarantee of rights and the freedoms that we enjoy, with limitations on what the government is allowed to do, combined with the free market system that they put into place, have produced the greatest freedom and prosperity known to man; a system that has enabled talented, hard-working individuals to use their talents to create wealth and the opportunity for prosperity for all Americans, and a standard of living unmatched in the world for all Americans, willing to do a fair day's work for a fair day's pay.

I'm reminded of that old saying that we have all heard many times before: You had better be careful what you wish for because you just might get it. Apparently, a lot of Americans wished/hoped for change a few years back. But change from what and to what? Did we ask for change from the system that has produced the greatest freedom and prosperity in the history of the world to a system that has already been tried many times throughout history and proved to produce neither? Apparently, a lot of politicians in Washington think so.

What Will Be Our Legacy?

Virtually all Americans, who are *not* individually extraordinary by world standards, enjoy freedoms and opportunities for prosperity that few others in the world do, and it has been provided to us by our Founding Fathers and from the sacrifices of our ancestors. My greatest fear is that we are not going to leave the same legacy for our descendants. Let's hope we wake up before it is too late.

The next time any of us are tempted to complain about the wages that we are making and the living standards that we have (especially if we are being compensated well above what the free market would establish), we need to ask ourselves what our labor and skill sets would be worth in another country, say, for instance, in Mexico or Honduras or Portugal or South Korea or the Philippines. Once we do this, we might be a little more diligent in trying to protect and enhance what we have been given, that which many around the world are willing to risk life and limb to obtain.

Chapter 1

What Would You Expect with a Parent Like GM?

I graduated from SMU in Dallas, Texas, in December of 1970 with a degree in Industrial Engineering and a strong desire to find a good job that would be professionally rewarding, as well as provide an opportunity for a good living for my family. Over the prior several months, I had been interviewing with a number of highly respected companies, large and small, manufacturing and nonmanufacturing, local and distant.

Growing up in East Texas and during my university years, I held a number of summer jobs, most of which gave me a pretty good idea of what I *didn't* want to do the rest of my life, and which kept me working hard in school. These jobs included working in the hay and watermelon fields, life guarding, working on the loading docks, driving a bread truck, painting, and jailer in the Dallas County jail. Not only did these jobs help keep me out of trouble and provide spending money, they helped teach me a valuable work ethic and an appreciation of a variety of work environments.

Based on my prior work experience and impressions I had throughout the interview process, I came to the conclusion that I wanted to work with a successful manufacturing company. I think this conclusion came pretty naturally, as it has been my observation over the years that many conservative people, such as I am, gravitate toward professions where they perceive that products and services of real value will be provided to society.

After interviewing with several companies that came on campus and receiving some very good offers, I had just about decided to take a job with the largest home builder in the Dallas area. Prior to accepting this offer, however, I did want to interview with three large, very prominent manufacturing companies that were located not too far from northeast Ohio where my family had settled following my father's recent transfer. Family has always been very important to me, and I decided that if one of these companies looked as professional as the Dallas home builder and made a comparable offer, proximity to my family would be the tiebreaker.

As it turned out, all three companies were quite impressive and made competitive offers. The most impressive of the three, however, was Packard Electric, which also happened to be located the closest to my parents' new home, only an hour away. The fact that I was escorted around Packard during my interview day by an executive who also had played major college football (Penn State), and who was also an engineering graduate, didn't hurt either. I knew that he would understand the hard work and sacrifices necessary to accomplish this feat, and that it would probably work to my advantage in this company. It also didn't hurt that Packard was a subsidiary of General Motors, which was the largest company in the world at that time. This undoubtedly meant that it was well managed and would be around forever, at least that is what I thought at the time. I was still very young and naïve.

But, in addition to these things, Packard Electric was a very impressive company. At that time, all of Packard's operations were in Warren, Ohio. (It didn't take long after signing on to realize how foolish this was as the entrenchment and inflexibility of the union severely restricted Packard's ability to run an efficient operation). As best I remember, Packard had over a million square feet of floor space under roof, most of it being manufacturing floor space. However, they also had a very impressive two-story engineering building in addition to the normal office buildings. When I saw the reverence paid to the Methods Lab and the importance placed on the Industrial Engineering Department, I was hooked.

Interview Process

In reflecting back on the interview process, there were a few things that I found interesting. For one thing, there was not a lot of consistency in the interview process from one company to the next. In some cases, the interviews were done in a very relaxed and informal manner, and, in other cases, the interviews were much more structured and formal. In some cases, the bulk of the interview centered on my work background, goals and aspirations, likes and dislikes, and how I could be an asset to the company.

In other cases, the interviewers seemed more interested in my family background and in just chewing the fat. I came to the conclusion that this was due, not only to the personality and attitudes of the interviewer, but also, in part, to the culture of the company, which can be a good clue as to whether the interviewee would be comfortable in that work environment.

I was also astonished to find that one of the companies in Ohio, with whom I interviewed, which happens to be one of the world's largest manufacturers of household products, had the audacity to give me, as well as the others that they were interviewing, a fairly comprehensive test, which was similar to an ACT or SAT test. I took the test and knew I had done well, but I quickly decided that this was a company for which I did not want to work. Apparently, they just didn't trust a degree with honors from a major university. If they didn't trust this, what else would they not trust? And, if they were primarily looking for people who would score well on tests, did they really have all of their priorities straight?

The last observation, which was kind of surprising at the time, is that each of the offers I received was within a few dollars of the others, and the benefits packages were very similar as well. This was the case in nonmanufacturing as well as manufacturing companies, and it also was true in union versus nonunion companies in which there may have been significant differences in the hourly wage rates and benefits. (Of course, this did not indicate what would happen to salaries and benefits going forward, but, in the case of Packard, based on studies that the company itself conducted over time, salaries and benefits for salaried employees were never out of line with industry norms. This is certainly not the case with the wages and benefits of union employees, which, in large measure, are responsible for the dire straights that Delphi, GM, and other unionized companies now face.)

Packard Electric Is Born

Packard Electric (not to be confused with Hewlett-Packard, which frequently occurred when I would tell people where I worked) was established in 1890 by the Packard brothers, James and William, in Warren, Ohio, in heavily industrialized northeast Ohio, near Youngstown, and not far from Akron and Cleveland. The new company manufactured incandescent lights.

Many of you, especially if you have been around for 55 or more years, are undoubtedly familiar with the Packard automobile. It so happens that the Packard brothers also founded the Packard Automobile Company as a subsidiary of Packard Electric. My dad owned a Packard convertible that he bought after returning home from the Great War, and this was the car my parents owned when they were married in 1946. My dad, for some nutty reason, such as "practicality," sold this car and did not keep it to hand down to me.

In 1932, it appears that General Motors was well into its strategy of vertical integration, i.e., acquiring or developing automotive components companies to supply its automotive assembly plants with the parts needed, rather than purchase those parts from outside vendors. By this time, Packard Electric had moved into automotive wiring, and the new acquisition began to supply General Motors with virtually all of its wiring harness needs for the next 70 or so years (see sidebar).

For some of you not familiar with automobile wiring harnesses, go to your car and lift up the hood. You will see wires bundled in black conduit or tape with individual wires plugged into connectors that are fastened to virtually all of the components in the engine. Everything electrical in the car runs off of 12-volt direct current supplied by the battery and which is distributed to the car by a network of wiring harnesses. The engine harness, and the harness that controls the front lights, are visible under the hood, and the harness that supplies power to the rear lights is generally at least partially visible in the trunk. Harnesses that provide power to the entire instrument panel and to the interior lights, as well as the harness that carries power from the front of the car to the back, are normally hidden under carpet, trim, paneling, or the dash.

I'm sure that at the time this strategy seemed like a good idea. It would provide GM with a dependable supply of needed parts, while giving them considerable control over all aspects of the supplier's business, in that course corrections or changes could be made as needed. It would allow GM to better control and protect new technology in order to potentially give it a competitive edge in the market place. And, as an added bonus, some of the vertically integrated companies also would have outside sales, which could add to the coffers.

But, what would be the long-term ramifications; did anyone ponder this?

Over the years, GM established a large number of these vertically integrated suppliers, companies of which virtually everyone has heard, such as Fisher Body, Guide Lamp, Delco Remy, Delco Electronics, Harrison Radiator, Saginaw Steering, etc. Starting no later than the late 70's, GM finally started realizing that the strategy, in fact, had created some very unpleasant, unintended consequences. However, by then there were 26 vertically integrated automotive parts companies within the GM corporation, all of which were highly noncompetitive, some more so than others.

Packard was actually in one of the better positions of all of the GM subsidiaries. In part, this is because Packard had had some pretty good management over the years. The fact that the entire Packard worldwide operation was located in Warren, Ohio, with 13,000 union workers earning UAW (United Auto Workers) wages was much more the fault of the parent company, GM, than it was the fault of Packard. (Packard was actually organized by the International Union of Electrical Workers or the IUE, but the UAW contracts actually dictated wages and benefits.) It just hadn't been "convenient" to decentralize up to now, despite the recommendations of Peter Drucker (writer, management consultant, and "social ecologist," 1909–2005) and others, and the dictates of good common sense (which, within the GM family, was not all that common based on my observations).

It is also true that Packard's products have much less a commodity nature than the parts of several of her sister companies. A wiring harness is not something that can just be purchased off of the shelf or out of a catalog.

The concept of standardized and interchangeable parts dates back to Eli Whitney and the cotton gin, and has been effectively used in the automobile industry dating back to Henry Ford. However, each wiring harness type for each different model requires a different wiring harness design. There is a tremendous amount of cooperative engineering work required between GM and the supplier to get wiring harnesses designed in a timely and effective manner. Even at that, wiring harnesses are the most susceptible of all automotive components to late changes. The pertinent phrase we used at Packard was "wires bend," but this is certainly not the case with almost any other car part.

Virtually any late change in a vehicle design will dictate a wiring harness change, either to accommodate a functional change or a routing change necessitated by something having to be moved, added, or deleted within the car. This gave Packard a real advantage in establishing a good (profitable) price. For starters, the true cost estimate for a wiring harness is more difficult to establish than the price for most other components; almost more like an art than a science, given the subtleties of wiring harness manufacturing. And, maybe more importantly, the frequent engineering changes provided many opportunities for repricing, which is what created the majority of the profit margin for Packard's products.

To be fair, GM wasn't the only car company to incorporate vertical integration. It also was practiced by others, most notably Ford (and it didn't work out too well for them either with Visteon), although not to the extent of GM. But, couldn't the car company executive have looked into a crystal ball and seen what the long-term consequences of this type of strategy might be? Apparently not, but I would contend that it shouldn't have been that difficult, and, in fact, what did occur over time should have been easy to predict.

The condition that existed at that time that seemed to make the decision to utilize vertical integration viable is that there was very little competition in the U.S. automobile market prior to the oil embargo of 1974. Europeans produced some nice cars, but their automobile makers were as heavily unionized as ours and were not a real threat to make big inroads into GM's approximate 50% share of the U.S. market, and Ford, with almost a 30% share, apparently wasn't too concerned either.

Japan was not yet a player in the U.S. market, and apparently U.S. auto executives didn't think they ever would be. Those of us who remember the 1950s and 1960s recall that Japanese products were thought of, at the time, as cheap and cheesy. I think a lot of auto executives thought (or maybe it was just hoped) that Japan never would get its act together and be able to make a product able to compete with the big three. I can remember several times during my early Packard career of hearing one executive or the other state that Japan would never be a serious threat to the U.S. automakers. They would say something like: "Sure, Japan may be able to make a decent little car, but they will never be able to make a suitable big and/or luxury car"; just the type of car preferred by the American car-buying public (try telling that to Lexus or Infiniti owners now).

Of course, the 1974 oil embargo changed a lot of that type of thinking (but certainly not all of it). Japan had a lot of small cars in production with excess capacity, and they were pretty good cars at that. With gas prices soaring, small fuel-efficient cars became the rage, and the U.S. makers had nothing with which to compete. The perception of Japanese products started changing right then, and the perception has only continued to improve, primarily in the consumer electronics and automotive markets where they have put most of their efforts.

The American auto industry quickly learned that the Japanese could, in fact, design and build a very high-quality car that Americans would be ever so happy to purchase, because their small cars had already changed the perception of Japanese vehicles in the minds of American consumers. However, even then, I remember hearing several different Packard executives state something like: "Sure, Japan can build a pretty good large car, but they will never be able to build a full-size SUV or pickup." (Okay, their little trucks might be pretty good, but we're talking about big trucks here.)

"That's where the money is and we will always dominate in that market." And, I'm sure very similar things were being said by GM executives. It's probably where Packard executives picked it up.

How Could GM (Packard's Parents) Make So Many Dumb Decisions?

I'm sure a lot of people are wondering how it is that GM (being, for most of the twentieth century, the world's largest company as well as one of the most profitable) made so many bad decisions and got into such a dire situation requiring a $50 billion bailout from the government for its very survival.

How could a company allow itself to get into a situation where it was buying most of its parts from subsidiary companies that paid UAW wages and benefits, when most of its competitors were buying parts from suppliers around the world that had labor costs only a fraction as high? How could a company get itself into a position where its subsidiary suppliers had their entire operations in one small geographical area? How could a company not have a single competitive small car available when the oil embargo hit, knowing (or at least it should have understood) that there would undoubtedly be large fluctuations in fuel prices over time, because the United States was no longer energy independent? How could a company lose its brand identity in the 1980s by making so many look-alike cars, while trying to sell them for vastly different prices to different clientele? And so on and so on.

Before giving my thoughts, let me quote from Jerry Flint (the late senior editor for *Forbes Magazine* who followed the car industry closely) from a speech he made in 2000 to GM Proving Ground engineers and technicians. "Everybody makes mistakes. But your management makes so many of them. The proof of their incompetence is in the number of mistakes. There is absolutely no reason to think that this will change. The same people who made the mistakes are still in charge, and they haven't admitted it.

"Listen carefully, you have management that doesn't know much about the American car business. It isn't that they are bad people or dumb people. I assume they are smart. They just don't know much about the American car business. ... How can anyone who knows something about the American car business, about cars, get to the top, or even the no. 2 position, of GM. I don't see a pathway up. Engineers don't count for anything anymore in this company as far as I can tell."

Flint goes on to talk about scores of major mistakes that GM had made in the prior decade or so, things such as the Saturn disaster, being constantly late to market, designing some horrible vehicles, not having affordable convertibles to match the competition, and not having any decent engines when at one time it was the leader in the industry, just to name a few.

I'm sure you get the picture. Unfortunately, most of us who worked for GM in the past could say nothing but "amen" to what he preached.

GM has had financial guys running the show almost from the beginning, and this makes it easier to understand many of the mistakes that GM management has made over the years. It is apparent that many of the decisions were made by bean counters based on short-term economics, but without a lot of consideration as to what the long-term consequences would be.

I think there is another reason why it was so easy to get short-term decisions accepted at GM, even though anyone with any degree of insight would understand that the long-term consequences could be very dangerous and damaging. The reason is that a sizeable share of all of GM executives' compensation is from bonuses, in large part based on short-term profitability and other short-term factors.

Remember my favorite saying from the introduction? "People will do what they are motivated to do, or the corollary, you will get the performance you motivate." When key individuals understand that a good part of their pay is based on the three-month profit and loss (P&L) statements, it is possible that they are going to make sure that they do not do anything to jeopardize short-term profitability, even at the expense of the long-term viability of the company. Human nature would suggest so, especially when you figure you have only got a few years to maximize your compensation before you are out the door.

I'm confident that is why most Japanese companies, which are greatly concerned about the long-term health of their companies, rarely pay bonuses (as opposed to profit sharing in which virtually everyone participates). They don't want employees making the wrong decisions based on what is in their best interests individually, as opposed to what is in the long-term best interests of the company. Japanese companies determine what the value of a job is, and then compensate their employees accordingly. Makes sense to me.

There are probably some fields where the paying of bonuses (or probably more accurately stated, the paying of commissions) is a motivating factor that would not have negative consequences for the long-term health of the company, and, therefore, may well be appropriate. This could apply to sales, for example, assuming that there was minimum risk that doing so would motivate someone to undermine others in the firm. I also could see it applying to a baseball player who was paid a bonus based on RBIs, home runs, runs scored, on-base percentage, earned run average, stolen base percentage etc., because achievements in these areas would enhance, not jeopardize, the team performance. However, paying bonuses strictly for the number of home runs hit or batters struck out could pose risks to the team performance. In the latter cases, a player could be motivated to pad individual stats at the expense of the team.

The bottom line is that I believe it is foolish for a company to tie a large part of an individual's compensation to a bonus that could provide motivation to sacrifice the long-term health of that company for personal gain. Do you think the United States would have the economic disaster on its hands it now does if the executives at Fanny Mae and Freddie Mac were not paid gigantic bonuses for cooking the books largely by making loans to unqualified clients? (In 2006, the Office of Federal Housing Enterprise Oversight [OFMEO] sued Fannie Mae Chief Executive Officer Franklin Raines, Chief Financial Officer Tim Howard, and Controller Leanne Spencer for $100 million for their part in an accounting scandal that included manipulations to reach quarterly earnings targets that would trigger hundreds of millions of dollars in bonuses. The executives eventually paid $31 million in fines.) Do you think executives at GM might have made different decisions if none of the executives had personal economic incentives to maximize short-term profitability?

As anyone who has had experience with a union will testify, it is not easy to stand up to organized labor. For a car company, a work stoppage can be very expensive and lead to the consumption of a lot of antacids. When U.S. car companies had very little international competition and they could sell virtually everything they made with a lot of pricing power (which meant they were very profitable), it was a lot less painful (in the short-term) to just cave in and give the union what it wanted. They could just pass the additional costs for labor on to the consumer.

Could the car company executives really have believed that there would be no day of reckoning? Could they have been that naïve? One would hope not, but they let things deteriorate to the point they are now. They were unbelievably shortsighted, and why? As Jerry Flint stated, "… incompetence could certainly be part of the problem," but the wrong kind of motivation could be another.

I think the same kind of factors were in play regarding the foolish policy of having most of the company's parts come from subsidiaries having their facilities concentrated in one location, which had to pay UAW wages and benefits. It would have been costly in the short run, and a lot of trouble, to decentralize and/or divest itself of its subsidiaries, but the writing had been on the wall for decades. When GM finally got around to consolidating its surviving, vertically integrated suppliers in 1994 and spinning them off in 1999, it was already several few decades too late. The die had already been cast.

Similar thinking had to have been behind GM's reluctance to design and build small, fuel-efficient, affordable cars, a market that Japan was starting to exploit even before the oil embargo. New cars are expensive to bring to market and the margins are not as great on small cars as on larger cars, so why build them? This had to be the thinking. Again, would the decision have been the same if there wasn't motivation to maximize short-term profitability, or if there had not been a financial guy making the final decision? I hope not.

It may sound like I have been a little hard on GM, but as a former employee, I believe I am being more than fair. I think John DeLorean, of DeLorean car fame and one of the former top executives at GM, who ran both the Pontiac and the Chevrolet divisions, would agree with me. He said in J. Patrick Wright's book, *On a Clear Day You Can See GM* (Wright Enterprises: Books, 1979), that "GM has an inept management populated by fawning sycophants." Part of the problem, he claims, is that "the executives who rise to the top are not the best managers, but the men most skillful in flattering their bosses." He said, "A system which puts emphasis on form and style and unwavering support for the decisions of the boss (*the new term for this type of employee is "empty suit"*) almost always loses its perspective about an executive's business competence." This was written in the late 1970s and I never heard anyone at GM disagree with his assertions, although, I will say I was not getting feedback from the "empty suits" to whom DeLorean was referring. I did, however, run into a number of them, as Packard assignments took me to the various assembly plants, and they were not very hard to identify.

Peter Drucker, the noted management guru, also had some strong advice for GM. He had been given access to GM in the early 1940s, and for two years he was allowed to attend every board meeting, interview employees, and analyze production and decision-making processes. At the end of the process he wrote the book *Concepts of the Corporation* (John Day Co., 1946). In it he suggested that GM reexamine a host of long-standing policies on customer relations, dealer relations, and employee relations, among other things. Drucker was also a big fan of decentralization and simplification and believed that corporations tend to produce too many products (*no comment needed here*), hire employees they don't need when a better solution would be outsourcing (i.e., limited or no vertical integration), and expand into economic sectors that they should avoid (such as EDS and Hughes). Obviously, GM ignored all of Drucker's counsel over the years, and he became persona non grata within GM for the rest of his days. It seems GM management doesn't like those who disagree with them, whether they are inside the company or out.

Just to confirm that Drucker was not completely off base, Jerry Flint made the following statement in the speech I referred to earlier: "Your company has the poorest supplier relations in the industry, and a reputation of mistreating suppliers, of trying to beat down their prices unfairly. If someone comes up with a great innovation, GM is the last company it will try to sell it to for these reasons. I have had CEOs of major suppliers say this. Yet, this is how your management does business."

Just to make sure it didn't play favorites, GM also disregarded the advice of Edwards Deming (American statistician and consultant), who gained fame with his work in Japan, especially with Toyota. He taught that companies should end the practice of awarding business on the basis of a price tag alone. Instead, he said the objective should be to minimize total cost. He advocated moving toward a single supplier for any one item, on a long-term relationship of loyalty and trust. (*I actually think having two suppliers, with a 70–30 or 60–40 volume mix, makes more sense, in order to minimize risk and create friendly competition, which is what Toyota often does, but Deming was spot on with the rest of his advice.*) GM often claims to have a great reverence for Toyota and the Toyota Production System, but I found that its implementation of Toyota's concepts was very selective, and when GM and her subsidiaries did try to implement Toyota concepts, they were usually misunderstood and misapplied. The debacle with Saturn is a good case in point. You know, the car that was going to beat the Japanese at their own game, but is now defunct (see sidebar).

I must make an insert here based on the recent high-visibility quality problems that Toyota has experienced since I initially wrote this chapter. No one doubts that Toyota has been a world-class manufacturer for the past several decades and it reached that point by adhering rigidly to several manufacturing principles, most of which are shared in Chapter 4. I don't think the jury is in yet on just what happened to create the problems we have heard about recently, but the Toyota chairman probably hit the nail on the head when he told Congress that they probably grew too fast. My translation: They got away from the basics that made them great. GM and Delphi know all about this phenomenon.

So, How Are Things Working Out for GM Today?

So, how are things going at GM today? If you must ask, not well. In fact, unless you have been living in a cave, you know that GM would be insolvent if not for government money ("taxpayers"), and, boy, isn't that a great situation in which to be. And I'm starting to see more of GM's historic shortsightedness. It is really a shame that GM did not go through a structured bankruptcy, which would have gotten it out from under oppressive labor agreements and would have allowed it to reorganize its business for long-term success. The government, which many of us are convinced cannot do anything right, is now in the position of being able to tell GM (and a lot of other companies foolish enough, or desperate enough, to accept government cash) how to run their businesses. If this doesn't make you shudder, nothing will.

The Obama administration decided that Rick Wagoner, who had been running GM since 2000, must go. Wagoner, by the way, was a financial guy. The problem is that he was replaced by Fritz Henderson, another financial guy, who was replaced by Ed Whitacre, Jr., who has now been replaced by Dan Ackerman—none of whom are car guys. I am again reminded of the definition of insanity: Doing the same things you have always done, the same way you have always done them, and expecting that the results are going to be different. It is entirely possible that Wagoner was not the right man for the job, but it would definitely not be known by Obama or anyone in his administration, which doesn't know the first thing about the automobile business. I would certainly come closer to trusting the GM board (which itself hasn't been too trustworthy lately) than I would come to trusting anyone in the government to determine the right course of action for any company, especially one like GM. GM has sold its soul to the devil, and I, for one, am not very optimistic about its future.

Getting back to GM's shortsightedness. Why was GM so reluctant to take the one course of action that it needed to take if it is to survive long term? And, that course of action was a standard bankruptcy, in which (among other things) the union contract would be voided and could be replaced by a more sane and reasonable contract.

Instead, they allowed themselves to be coerced into a government-controlled bankruptcy, in which the government will call the shots (or at least have veto power), regardless of Obama's protests to the contrary.

Obama has protected a major constituency and contributor, at the expense of the American people. The union and its workers have come out of this debacle virtually unscathed i.e., their pensions and health care benefits have been protected and the union health fund was given 17.5% of GM shares. Whole bond holders took a bath, losing about 60% of their investments. (even while being largely responsible for the mess), while non-executive salary workers and retirees and the tax-paying American public will take a bath (even though these groups are virtually blameless for the current problems in which GM finds itself).

But, why GM's reluctance to enter into a standard bankruptcy instead of a government orchestrated bailout?

As long as the U.S. manufacturers have strong unions and their competition (even those manufacturing in the United States) do not, the U.S. manufacturers will never be able to compete on a level playing field, even with the home-field advantage. I really do believe most Americans would rather buy a car from a U.S. manufacturer than a foreign one, even a transplanted foreign one, all other things being equal).

Unions have had no success in organizing these foreign-owned plants, and they have been hard at it for 20 years. Part of this is undoubtedly due to the policy by the transplants of paying fairly competitive wages and benefits to their employees, but a good deal of the union's lack of success has to be attributed to the good sense of the workers. For decades, they have seen what unions have done to so many U.S. industries, most notable the steel and automobile industries (without even considering how teachers' unions have gravely injured public education, and how other unions of government and of service employees have inconvenienced and, in some cases, endangered everyday life as evidenced by the many instances of Blue Flu and strikes by firefighters, air traffic controllers, sanitation workers, etc.). Why should an employee, who has a great job in which he/she is treated fairly and paid very well in the market place, pay several hundred dollars a year to a union who might very well lead (or drive) that company to destruction?

These workers have shown, at least to date, that they have a lot more foresight than either U.S. automobile executives or union leaders (who, if anything, are even more shortsighted than the auto executives). One can certainly lay blame at the feet of the auto execs for caving in to the union and agreeing to compensation packages that were not sustainable. However, as much or more blame can be laid at the feet of union leaders who were unyielding in their unreasonable demands. I lived through one strong-armed strike while employed at Packard, which lasted only one day, and that was more than enough to last me a lifetime.

The bottom line is that the compensation packages for current workers at the big three, versus the foreign-owned U.S. assembly plants, is not gigantic. According to Manufacturing.net, the hourly wages for the big three are around $27 per hour with no profit sharing. With profit sharing, Toyota's hourly rate is a little higher than this number. The other foreign-owned plants have a little lower rate. When all benefits are factored in, the fully burdened hourly rate for *current* employees is about $48 for Toyota (somewhat lower for the other transplants) versus somewhere in the area of $55 to $60 for the Big 3.

One of the critical problems that the Big 3 face is that they claim a fully burdened cost per hour of labor of $60 or more per hour. This difference is due to the impact of retirees, who have been promised generous benefits after 30 years of service. The pension plans have been funded (for the most part, although the recent crash of the market has certainly hurt the funding level), so this is not a big part of the problem. The problem is that healthcare was promised to retirees, both hourly and salary, until the age of 65, with some help thereafter, and little, if any, of this commitment has been funded. The transplants do not suffer this cost because they have not been in operation long enough to have retirees, and I am confident they will not make long-term commitments to retirees that they cannot possibly maintain, if for no other reason than the lesson learned from the Big 3.

Another very big problem that GM has, as well as Ford and Chrysler, is the negative impact the union has on productivity and quality. Anyone who has worked in a union shop or has had dealing with a union knows this to be true. It would be one thing if good people were being abused in the workforce (as once was the case in this country before unions), but this is no longer the case and the union has outlived its usefulness. Today, due to intense competition (that did not exist in the distant past), a company or organization is not going to be able to stay in business if it does not treat its employees both fairly and well.

The UAW has been very shortsighted, not only in its excessive compensation and retirement demands, but also in its policy of fighting for and protecting individuals that by all rights should be fired. Most employees who know that they won't lose their jobs regardless of how they perform, are not going to perform as well as those who know they could lose their jobs if they mess up badly enough. When a company has to live with an employee that by all rights should be fired, it is certainly discouraging to those workers giving their best. This clearly undermines plant performance, as does the UAW policy (unfortunately, nearly all unions have this policy) of fighting for job classification restrictions that result in unnecessary jobs that make the company less competitive, but swell union ranks (and union dues). Why have union leaders, much like management, not been able to see past the end of their noses and try to project what the long-term consequences of their policies and decisions might be?

So, again, why was GM so reluctant to go into a standard bankruptcy, especially since the government indicated a willingness to make a commitment to consumers of GM products that it would help ensure warranty and service? Maybe it is just more shortsightedness on the part of management, or maybe it was the heavy handedness of the government. If GM is to survive, it will take more than just a restructuring of the company. As much or more than anything else, it will require a gigantic restructuring of the agreements with the union, and this is just not part of the agenda of the government-controlled bankruptcy. The very minor concessions and "sacrifices" made by the union are just not going to make much of a difference in the long run.

We will just have to wait and see what ultimately shakes out regarding GM's emergence from the government-controlled bankruptcy. So far, nothing has been done to fix the multitude of union-originated problems, but a lot has been done to disenfranchise salary workers, retirees, and the American public at large. Even disregarding these problems, GM has a long way to go to compete with Toyota and other international car companies.

How Does GM Compete with Toyota and the Vaunted Toyota Production System?

I'm going to spend a lot of time in this book talking about the Toyota Production System with regards to Delphi Packard, with which I am very familiar, so I will not talk a lot about it now with regards to GM. However, there are a few things I would like to mention.

For one thing, as best I can tell, there are very few, if any, people within the GM organization, now or in the past, who have really understood what the Toyota Production System (TPS) really is. This is so, even after countless visits, studies, white papers, and even a joint venture with Toyota. Toyota has used a lot of tools over the years to solve problems; tools such as quality circles, kanban cards, andon cords, various poka-yokes, stockless manufacturing, etc. The assumption of most GM managers (and also Delphi managers) is that these tools are an integral part of the TPS and a large part of their success.

The truth is that Toyota does not consider any of these tools as fundamental to the TPS. Toyota uses them merely as temporary responses to specific problems that will serve until a better approach is found or conditions change (as pointed out by Steven Spear and Kent Bowen in a paper entitled "Decoding the DNA of the TPS" from the October 1999 *Harvard Business Review*).

However, GM and Delphi seem to consider each of these tools as "an end in itself," rather than as "a means to an end." This is a tremendous distinction, and a very big blunder.

In the late 1970s, Saturn was the new GM car company that was going to revolutionize American car building. All of the old ways of doing things within GM were going to be replaced by doing things the Toyota way, including the team concept, the idea of getting everyone's buy-in to decisions, and the use of team leaders. Unfortunately, the Saturn team didn't have a clue as to what Toyota was really doing to make itself successful. Edwards Deming said it well: "I think that people here expect miracles. American management thinks that they can just copy from Japan, but they don't know what to copy." How true this is.

Another statement of his is also pertinent: "Information is not knowledge. The world is drowning in information, but is slow in acquisition of knowledge. There is no substitute for knowledge." GM (as well as Delphi) has a lot of information. They are just a tad short on knowledge, at least as it pertains to the TPS.

The result is (and this is not an overstatement) that the Saturn experience has been an unmitigated disaster. All of GM was and is in big trouble, but no part any more so than Saturn. It's true that, for the most part, the product line has not been very exciting, but the operation of the factory has been a colossal failure. Much of what they tried to put into place, with so much fanfare, failed miserably and fell by the wayside.

The real problem with GM is captured very succinctly by John McElroy in an article entitled "Is GM Better or Ford Worse," written in the July 1, 2002, *Ward's Auto World*. He writes: "American companies don't seem to have the discipline or attention to detail needed to grind out consistent annual improvements. They can do spectacularly well for a while, but ultimately succumb to distractions when management gets bored with consistency and starts dreaming of breakthrough programs that will leapfrog the competition. Toyota and Honda still lead everyone else because they stick to what works and constantly improve it. They are not smarter. They are not better. They are simply more consistent."

How true these words are, and they are especially true for Delphi, as I will demonstrate later. For now, humor me and let me give one sports analogy. If you analyze virtually all of the great sports teams throughout history, you find one thing that is almost universally common, and that is that they have all achieved an unwavering adherence to the basics, whatever the sport. More often than not, it is not the team with the most talent that wins, but the team which accomplishes the basics most effectively.

I think back to the old Green Bay Packer teams under Vince Lombardi. Lombardi took over a team of losers and turned them into perennial champs. He didn't do it by gimmickry, but rather, he did it by getting his team to block, run, throw, catch, and tackle better than anyone else in the league. Other teams tried to copy his plays and formations, but none were very successful. It wasn't the plays, it wasn't the formations, it wasn't even necessarily the players. It was the team's success in doing the little (but very important) things right. Teams that have taken focus off of the basics and put a lot of energy into trick plays or other gimmicks just never have been very successful.

The good news for GM is that even with the loss of productivity due to the union and the fact that GM has not accomplished the basics at all well, there is not a huge gap between GM and Toyota and the other foreign manufacturers in the hours required to build a car. In fact, back in the 1990s Chrysler was thought to be the most efficient of all the car producers, based on hours per car. These numbers are a little hard to ascertain accurately due to differences in cars, options offered, and internal plant information, but most of what I have seen shows a gap of less than 10%. With reasonable union concessions on job classifications and work rules and with more concentration on the basics, this could be made up rather quickly. There are also a few key concepts of the TPS that, if understood and adopted, could prove beneficial. Whether GM and the union have the wisdom and dedication to make it happen is another issue.

Chapter 2

We Have to Do Something, Even if It's Wrong

I started with Packard Electric on January 19, 1971, on a bitterly cold day. If the weather had been like this during my Packard interview I might very well have ended up in Dallas building homes, instead of in Warren, Ohio, building wiring harnesses. Who knows how that might have worked out. But, for one thing, I wouldn't be writing this book, which I believe can help manufacturing companies, especially U.S. companies, become more competitive in this world economy.

When you have a large family, you think a lot about what the country and the world will be like for them. Will they have the same opportunities that we had? There is much to be concerned about, as the direction our country is taking drifts us farther and farther from the principles our Founding Fathers put into place. But, regarding our manufacturing competitiveness, I believe we can regain our once dominant position in the world if we learn from our mistakes and start implementing the concepts of "Intelligent Manufacturing" that will be introduced in this book.

On my first morning with Packard, I checked in with Personnel and went through the obligatory orientation, signed all the appropriate papers, and then went across town to the Engineering Building to start my new job. I was greeted by my new boss and several of my new workmates. I was then taken on a more detailed tour of the manufacturing facilities on both sides of town.

On the side of town adjacent to the Engineering Building was the cable-making operation and all of the wiring harness manufacturing facilities. I was even more impressed than I had been during my abbreviated tour during the interview process. The plants were relatively new, well laid out and organized, and the housekeeping was quite good. It also was very noticeable that all of the employees seemed to be busy doing constructive things, and there were no gangs of workers milling around. I had been in car assembly plants and other operations that had unions, and this was usually anything but the case (on one visit to an assembly plant, I even witnessed one group of hourly workers engaged in a card game during working hours, not to mention a few that were sleeping). The technology was not overly complex. In fact, the manufacturing processes were quite straightforward, but, it was obvious that the equipment was very functional and well maintained.

After my tour of these facilities, I was taken back to the other side of town to the factories near the main offices. These were Packard's original factories that had been in place since the early days of the company. The buildings had the architecture and appearance of early twentieth-century buildings, but they also were well maintained. These buildings held the component-making operations of Packard as well as the ignition cable manufacturing operation.

Whereas the newer facilities on the other side of town contained Packard's highly labor-intensive manufacturing (including the cable- and terminal-making areas), the buildings on this side of town held most of the highly capital-intensive manufacturing. Packard manufactured ignition cable and ignition cable sets, plastic components, vinyl tape, and rubber parts of its own design in these facilities, which were used on virtually all of the wiring harnesses Packard Product engineers designed (in conjunction with GM engineers) and Packard manufactured.

These facilities were very impressive from the standpoint of the enormous investment that had been made in tooling and equipment, but much less so from the standpoint of layout, housekeeping, and the apparent productivity of the employees. Part of this was undoubtedly due to the age of the facilities, but there was a difference you could detect. As I later discovered, this was primarily the result of two factors: the older Packard employees, with greater seniority, worked in these facilities and the productivity controls in these plants were not as good as those in the newer plants, to some degree due to the nature of the work performed.

Packard billed itself as a full service provider, which, in fact, it was. It designed all of the wiring harnesses required by GM on all of its vehicles.

It designed all of the necessary components and it manufactured virtually all of these components. Packard purchased almost all of the raw materials it needed (copper, various plastics, metal coils for terminals, etc.) and then produced the rest of what it needed to get wiring harnesses built and shipped to the various GM assembly plants.

The problem is that Packard only had one customer for all practical purposes: GM, so it was not subject to the intense competition that an independent company would face. This certainly thwarted Packard's development to some degree, which became very apparent when it was spun off by GM in 1999, along with the other Delphi companies.

Packard Does a Really Smart Thing Next

While the wide-aisle tour I was given during my interview process had taken only a couple of hours, it took the entire day to take this more in-depth tour through the facilities, and I'm pretty sure that I walked five miles during the process. Other than being a little foot weary, I was very excited to get started the next day on my real job. The next day I showed up bright and early for work and received a little bit of a surprise.

Instead of receiving a little bit of training, being given an assignment, and being turned loose on an unsuspecting plant, Packard did something really smart, something which they did for all of their new hire industrial engineers. Before being given any assignment or being asked to do anything in the plant, Packard put all of its new engineers through a very intensive two-week training program.

What a lot of companies do is assign one of its current employees to act as a big brother or sister to a new hire. Packard did this also, but, if this is all that is done, there are risks that some critical knowledge will never be properly passed on to or acquired by the new hire, along with the risk that the new hire may well inherit many of the bad habits, inadequacies, and/or prejudices of the big brother or sister.

The training program that Packard provided to its new industrial engineers was very well organized and had obviously evolved over a long period of time (see sidebar).

Other departments within Packard, most notably within other engineering disciplines, had some semblance of training programs, but nothing to compare to Industrial Engineering's training program. In fact, some key employees from other disciplines, such as Quality Engineers, Product Engineers, Materials and Production Control Engineers, and Manufacturing Supervisors were put through an abbreviated IE training program, as it gave them a much better understanding of the products Packard built, how they were built, and the interaction between all of the various departments within Packard.

Every single new IE went through this excellent training program before being allowed into the plant to mess something up before he/she knew what they were doing. And, a two-week training program was scheduled for each new engineer, even if he/she was to be the only participant. In fact, there was usually only one participant, which provided sufficient individual attention to ensure proper knowledge transfer.

The first week was comprised of a number of 2- to 4-hour sessions taught by practicing Industrial Engineers, who taught critical tools, concepts, and skills necessary to be effective in the various tasks required of all IEs. The second week of training was conducted in the Methods Lab; and, during the week, an actual detailed production plan was developed for a small wiring harness. I cannot over emphasize the importance of preplanning to Packard, which I will discuss in detail in Chapter 7.

Having all of the training conducted by practicing Industrial Engineers helped the new engineer-in-training develop bonds with his/her new workmates very quickly; and it was a great tool for helping the older engineers stay current with their skills and hone their training capabilities (IEs conducted a lot of training within Packard, much of it involving hourly workers). It also ensured that the reality of the shop floor was being taught, as opposed to theory (which often doesn't hold water) that is taught by those who may have observed, but never really done the job.

As a side note, one of the interesting things I observed over the years as Packard started listening more and more to consultants and various "experts" is that few, if any, of them (the consultants or experts, that is) had any real manufacturing experience (most were academics with lots of classroom theories), and none had any experience in the wiring harness business. There are differences in businesses and something that will work well in one will not necessarily work well in another. It was always amazing to me that Packard management placed more credence in the opinions of individuals (mostly academics) who had no experience in the wiring harness business than they did in their own opinions based on their real-world experiences, or in the opinions of long-time employees who had years of experience in the real world of wiring harness manufacturing.

Where Did the Packard Training Program Originate?

I was so impressed by the training program (which I will discuss more thoroughly at the end of this chapter) that I wanted to know much more about it. Most Packard employees knew about the Packard brothers and the early history of the company. However, as I inquired about the origin of the Packard IE Training Program, no one seemed to have a clue where it had come from or who had developed it. I didn't worry too much about it at the time, but I always wondered who the Packard pioneers were in the area of Industrial Engineering who had done such an outstanding job of developing the Methods Lab and the preplanning concepts, along with introducing other IE tools that proved so very effective.

I never was able to find out who, if anyone, in Packard history could be credited with the tools and concepts that had been developed. Based on my additional research and study, I came to the conclusion that the development of these tools and concepts were a collaborative effort on the part of a lot of good engineers who developed them over a relatively long period of time.

However, there is one thing that is very clear—much of what was contained in the Packard IE training program was taken liberally from two pioneers in Lean Manufacturing. Yes, I know that some of you probably have the same knee jerk reaction I do when you hear the term Lean Manufacturing or Synchronous Manufacturing or the like. After all, this was the manufacturing system that was going to make our manufacturing companies lean, mean, fighting machines; but, in fact, it made many of them less productive as focus was taken away from the basics and placed on tools that, in many cases, are not particularly helpful, even if applied appropriately.

Concepts of Lean Manufacturing (although it wasn't called that at the time) date back well over a century. Some say Lean Manufacturing started with Eli Whitney and the concept of interchangeable parts; others say it started with Ben Franklin and his contention that avoiding unnecessary cost could be more profitable than increasing sales. If Lean Manufacturing is defined as just using good common sense to make the manufacturing of something easier and less costly, examples undoubtedly go back to the Stone Age.

But, for the purposes of determining who fathered the lion's share of Lean concepts successfully developed and implemented by Packard Electric, we can look to Frederick Winslow Taylor and Frank Bunker Gilbreth, Sr. As I have researched these men, it is clear that their work and their concepts had more influence on Packard's Industrial Engineering prowess than anyone else's (although Henry Ford would be a close third, with his invention of the assembly line [or paced operations], which were prevalent throughout Packard's final assembly manufacturing). While these two men were different in many ways (e.g., Gilbreth was the father of 12 while Taylor was the father of 2 adopted orphans; Gilbreth had no formal education beyond high school while Taylor had a prelaw degree prior to changing his vocation, due to failing eyesight, and eventually obtained a mechanical engineering degree), they were very similar in other ways.

Both were born in the later half of the nineteenth century, Taylor in 1856 and Gilbreth in 1868, and both died fairly young: Taylor at 59 from pneumonia and Gilbreth at 55 from heart failure. But, most importantly, both had real-world experience in a manufacturing environment. Taylor began work after the 1873 depression as an industrial apprentice patternmaker and then later worked as a machine shop laborer at a Midvale Steel Works. He was later promoted to gang-boss, foreman, research director, and, finally, to chief engineer. Gilbreth started as a brick layer and later worked as a building contractor.

Both of these men, through their own personal experiences, knew that there were tremendous inadequacies in the way that work was performed in their day, and both decided to help do something about it. Taylor concentrated his efforts in the area of standardized work (which is one of the four critical rules of the Toyota Production System developed almost a century later), and he is the first man in recorded history who deemed work deserving of systematic observation and study. Prior to Taylor, all work was pretty much done rule-of-thumb. He created Planning Departments, staffed them with engineers, and then gave them responsibility to:

1. Develop scientific methods for doing work. (Absolutely critical, and very much incorporated into Packard's DNA in the early days, and one of the four critical rules of the Toyota Production System.)
2. Establish goals for productivity. (Deming was very much opposed to this concept almost a century later, but then again, Deming never had any real manufacturing experience; he was an academic with theories, some of which were very good and some of which weren't. He was dead wrong about this one.)
3. Establish systems of rewards for meeting goals. (Packard wisely adopted the concept of a fair day's work as opposed to incentive goals, meaning that standards were established based on what a "normal" operator working at a 100% effectivity, in regards to pace and work skills, should be able to accomplish. Rewards for meeting goals were mainly in the form of praise [and maybe an occasional donut], more desirable

assignments, opportunities to advance within the company, etc. Establishing piece-work goals or incentive goals and paying extra for exceeding these goals would never have worked within Packard and would have been a disaster. In fact, incentive goals have not worked very well where tried. I cannot imagine the nightmare of trying to establish baseline goals, when the employees and union are going to do everything possible to force them to the lowest level possible. And then, how do you properly plan your workforce when you are not really sure what your output will be?)
4. Train the personnel in how to use the methods and thereby meet the goals. (An absolute must and also very much a part of the Toyota Production System and standard operating procedure [SOP] at Packard.)

The accomplishment of these four responsibilities was a very high priority at Packard Electric. Taylor's concept of a Planning Department became the Industrial Engineering Department at Packard, and Packard's laboratory for accomplishing all four of these responsibilities became the Methods Lab.

Whereas Taylor concentrated on basic work standardization and the use of the stopwatch to establish standards against which performance could be measured, Gilbreth concentrated his efforts more on workplace layout and on individual movements required to perform work, so that problem areas could be identified and improvements could be made to make work physically less demanding and less time consuming (remember that he started out as a brick layer). Gilbreth used a motion picture camera to identify and time 18 basic motions of the hand and he used this knowledge to make processes more efficient by reducing the motions involved. He taught that all aspects of the workplace should be constantly questioned and improvements constantly adopted (also tenants of the Toyota Production System [TPS] and standard operating procedures at Packard).

Gilbreth was also was a believer in the concept that there was usually one best way to accomplish a task and this way should be found and implemented. If you think about it, this actually contradicts the concept of continuous improvement, but my experience is that it is true. I frequently remarked to workmates in the 1990s, after Packard had experienced two decades of productivity digression (referred to as "deprovement," or the opposite of improvement, at Packard) that, if we had just maintained the level of labor productivity we had in the early 1970s, we would be the most productive wiring harness producer in the world (although assuredly not the most competitive, due to high labor costs and other costly burdens).

To make another sports analogy, do you think the fact that Super Bowl champions or NBA Finals champions rarely repeat is due to the fact that the following year's champion is superior to the prior year champion? Or, do you believe it is most generally because the prior year's champion did not maintain the same level of performance that they were able to achieve during their championship year? My experience is that it is almost always the latter and not the former. If a champion was able to maintain its high level of performance over the long haul, we would see sports dynasties lasting decades, not just one-year wonders. The Boston Celtics of the 1950s and 1960s and the UCLA Bruins of the 1960s and 1970s were rare exceptions. Yes, they had great players, but so did a lot of other teams. Their secret was being consistently great at doing the little things (the basics) right.

I believe this is also the reason for Toyota's success. They have kept their eye on the ball and have not gotten distracted by the latest fads. It's not the tools they use (or gimmicks, if you will), which are not even considered part of the TPS. It is the few basic rules and the philosophy they have embedded into their core. They are content with consistency and small improvements and, thus, are able to outlast their competitors.

Supposedly, in the 1990s, Chrysler was the most productive car manufacturer in regards to hours used to produce a vehicle; today they are one of the worst.

Toyota just keeps plugging along, content with small improvements while maintaining what they have achieved and not backsliding. Remember again what John McElroy wrote in the July 1, 2002 issue of *Auto World*: "American companies don't seem to have the discipline or attention to detail needed to grind out consistent annual improvements. They can do spectacularly well for a while, but ultimately succumb to distractions when management gets bored with consistency and starts dreaming of breakthrough programs that will leapfrog the competition. Toyota and Honda still lead everyone else because they stick to what works and constantly improve it. They're not smarter. They're not better. They're simply more consistent."

It seems to me that Taylor and Gilbreth should have been friends and collaborators instead of the apparent antagonists that they were. Their work was really two sides of the same coin, and their combined work would eventually lead to Time and Motion Study, of which every industrial engineer is familiar. Nevertheless, Packard borrowed heavily from both of these pioneering giants in Lean Manufacturing. Taylor's focus on standardized work, the use of the scientific method, and the use of labor studies were very prevalent in the Packard IE training program (and in the three-week IE training program that I resurrected and augmented in the 1990s, the outline of which is found at the end of this chapter). Gilbreth's work was likewise very evident in the sessions on Ineffective Worker Movements, workplace design, ergonomics, and Act Breakdown. The work of both of these men, along with the contributions of Henry Ford, were melded together, along with some outstanding work on the part of some unknown Packard engineers, to create the extremely effective Packard Methods Lab.

There is no doubt in my mind that when I started with Packard these tools and concepts were used to enable the company to install manufacturing systems for wiring harnesses that were the best in the world, and which led to the smoothest model year startups with the best quality and the fewest problems among all wiring harness manufacturers. I have observed virtually all of the operations of Packard's competitors around the world, and I can say this with a great deal of confidence. Unfortunately for Packard, that was then and this is now. Packard has digressed in many ways, while undoubtedly many competitors have improved. The result is that Packard has a long way to go to become truly competitive, even as it has come out of bankruptcy and has shed much of its prior financial burden.

Good News and Bad News about Packard's Management

When I began working at Packard in the early 1970s, it was apparent that the company had had many good managers over the years, and the current crop of managers also was very effective (for the most part). My belief is that this was the case primarily due to one philosophy, which to my knowledge was never written down or formally stated, but which was practiced religiously. That philosophy was that no one should be promoted to an executive level position within Packard's operations areas unless he/she had experience in both manufacturing and engineering, preferably Industrial and Methods Engineering. This practice would ensure that Packard executives had a good understanding of the business and a good grasp of reality.

While this policy was very helpful in keeping unknowledgeable people out of key operations positions, it was not infallible. My experience is that it makes a good deal of difference at what level this experience is gained. If it is gained early in a career at the lower levels of Preplan/Industrial engineer and Manufacturing foreman/supervisor, the experience is very real and very beneficial. If the experience is gained at higher levels, the benefit is diminished to some extent as job responsibilities start becoming less real world and more theoretical.

The other benefit of having perceived high performers gain this experience at lower levels is that it is much more difficult to be an "empty suit" (i.e., all show and no go) because there are very clear job tasks and assignments that must be accomplished or it will become quite obvious that they weren't. At higher levels, empty suits can more easily get away with style and form, as opposed to substance; as the subordinates get things done, the "suit" is able to spend most of the time looking good and currying favor with the bosses.

One problem that did occur fairly often at Packard was that some perceived high performers, who came from the right schools or had the right pedigree, were put on career paths before they ever set foot in the door. These career paths, even when they passed through engineering and manufacturing, were generally laid out in such a way as to protect the golden prospect from failure (as well as his sponsor) by not allowing enough time in any area to truly experience success or failure. This led to some very talented people having a distorted view of reality and, undoubtedly, a feeling of inadequacy in never seeing the results of their efforts. The joke around Packard was that you could pick out a golden boy by the short time he spent in any one area before being sent off to the next (in some kind of a grandiose training program), and by the need to send in a team to fix the problems he left behind after each move.

A lot of what I have just described undoubtedly exists in many other companies, maybe most, as many of you reading this will recognize. However, with the philosophy of Packard that strong performers in the operations area should have both Industrial Engineering and Manufacturing supervisory experience, preferably at the lower levels, prior to being promoted to executive positions, it did minimize the number of really bad high-level managers who could single-handedly create a catastrophe.

From the day I started at Packard, cost reduction had always been a major focus, not only of Industrial Engineering, but of the entire operation. The whole purpose-of-being for the outstanding Methods Lab was to design, develop, implement, start up, and debug the very best manufacturing system possible for each of Packard's products. But, Packard's management knew that other steps would be needed if Packard was to remain a viable company, mainly because the cost of labor was so much higher than what competitors were paying for their labor in their nonunion factories. There was just no way of gaining sufficient superiority in productivity to offset the gigantic difference in labor costs.

This strategy was two-fold. The first part involved moving some of the existing business from Warren, Ohio, to low-wage rate areas (all they had to do was get the union's blessing—no problem, right?). The second part was to move the lion's share of all new business to low-wage rate areas, i.e., any new GM vehicles or additional business due to increased volumes or increased content within current products, which occurred annually as additional features were added to cars to keep them with or ahead of the competition.

A Move to the Warm South

Moving the company south was undoubtedly a necessary strategy and one that, if not undertaken, would have doomed Packard much earlier than what ultimately happened. The first foray was into Mississippi in the mid-1970s. The first plant was supported by about 50 former Warren salaried employees and it covered all areas of operations and administration. It was started up as a nonunion plant with wages and benefits well below those of the UAW and IUE, but still very good for that area of Mississippi and people lined up for blocks when openings were posted. Early on, there was some rough going, primarily because the decision had been made to make some fundamental changes to the way Packard had traditionally done things, particularly as to how the wire cutting and the lead preparation departments would be organized and the work performed.

The problems created were so severe that teams of support personnel were rushed to Mississippi to help out, and there was still a lot of concern that GM schedules would be missed, which was almost a capital crime. Due to some Herculean efforts, no schedules were missed, but there were a lot of valuable lessons learned (or at least there should have been). The systems were changed back to what was common practice in Warren and, from then on, things ran fairly smoothly.

The lesson that should have been learned, but unfortunately wasn't (as I will point out through other examples in this book), is that a manufacturing process or system should never be changed until there has been extensive testing and study to ascertain that, in fact, the change will be beneficial. And, if it is, it should be duplicated throughout the entire operation. This is something that is very much a part of the Toyota Production System today, to the extent that you could say it is written in blood. Nothing much was known about the TPS back in the mid-1970s. Who knows, maybe in the mid-1970s it wasn't even part of the TPS, but you would think that common sense would dictate that you should never introduce a new manufacturing concept or system unless it has been thoroughly tested and proven to be an enhancement of what is currently being done. To not do so could lead to irrevocable harm and severe damage to a company's finances and reputation.

The Mississippi Operations performed relatively well over the first few years, and Packard did save a lot of money versus what it would have cost to have expanded operations in Warren, not to mention the risk reduction that was gained by manufacturing in more than one location. During this time period, GM also built a couple of assembly plants in the South to try to get away from the union and the high-cost operations it had in the Midwest. As you can imagine, it did not take long for the UAW and the IUE to get interested in organizing these new plants and they started a strong effort to do just that. GM and Packard realized that having a union within their facilities represented a lot more negatives in trying to run efficient, cost-effective plants than just the higher cost for labor and benefits.

So, in an effort to try to avoid unionization, a decision was made in the late 1970s to extend union wages and benefits to all workers in the nonunionized plants. This ploy worked for a while for GM, and for a little longer for Packard's Mississippi Operations, but eventually all of these southern factories that had been set up as nonunion plants invited the union in, and that sealed their fate. No additional expansions were made in Mississippi after the first three plants were built, which had all been built before Packard's Mississippi Operations became unionized and, therefore, noncompetitive. There were no attempts to build other plants in the South, as it became apparent the union could not be kept out for more than a few years, at best. Another strategy was needed.

What We Need Are Suppliers

This led to two additional strategies by Packard management in the late 1970s to try to offset the high cost of manufacturing in Warren and to avoid union problems. The first strategy was one that had some short-term upside potential, but I would contend any Packard executive should have known it would have some very dangerous long-term risks, which would make the strategy very questionable. (I can say this because I did tell this to anyone who would listen, in my own politically incorrect way.) The problem was that this was what the big bosses wanted to do, so who was going to argue with them. We didn't have too many totally empty suits in positions of responsibility within

Packard at that time, but we did have some very lightweight ones, i.e., they did a pretty good job in some respects, but, in no way were they going to jeopardize their careers by disagreeing openly with their bosses (see sidebar).

A funny, outlandish, but true story: Packard always spells the word employee with only one "e" in all of its publications. When I asked why, I was told it was to save ink and space, since the word employee was used so frequently in company literature. They didn't have an answer when I queried that, if that was the case, why commonly used abbreviations were not used as well. This was met with profound silence. I suspect the real answer is that one of the bosses, back in time, spelled the word with only one "e" on some important document, and no one had the nerve to tell him he had misspelled the word.

This new strategy was to establish a supplier network to which Packard would outsource certain business. The suppliers would produce the new products early on with a lot of costly support from Packard, and then sell them back to Packard. Because they would be paying wages and providing benefits well below union scale, the theory was that we could both profit from this arrangement (so long as Packard could maintain the current prices with GM once it discovered the arrangement). I suspect that Packard did make a few bucks from the U.S. supplier program, but not nearly as much as claimed, because I don't think we ever did a very good job of capturing our costs of supporting this effort.

I suspect that some of you can guess what some of the potential risks from this arrangement might be, and I'm wondering that, if you can see some of the pitfalls, why was it not obvious to Packard executives? For one thing, Packard would be on the hook for any quality or delivery problems that were created by any of the suppliers. For another, GM would undoubtedly find out about the arrangement and ask for better pricing. They might even wonder: Why do we need Packard to be the middleman, since we (GM) are capable of designing our own wiring harnesses without Packard's help? Finally, is the practice of creating potential long-term competitors, for the sake of some short-term gain, a good idea? I also should add that Packard lost a lot of good engineers to these suppliers during the process.

I would contend that the U.S. Supplier Program was doomed from the start and should never have been undertaken, but at least there was some short-term gain for Packard. This is something that cannot be said about Packard's implementation of the same concept in Mexico. (I'm getting a little ahead of myself, but this is a good time to mention this.) After we started up operations in Mexico in 1978, Packard decided to implement a similar supplier strategy in Mexico, starting in 1981. This strategy made absolutely no sense to me and I expressed this very strongly. It may seem somewhat strange in that I was tasked with the engineering responsibilities to make this happen, but I was finally told to shut up, accept it, and get it done—which I did.

We started up two suppliers very successfully, at least from a quality and delivery standpoint, and some will say that it did take some of the pressure off of Packard's Mexican Operations. However, we lost a lot more than we gained, and this could have been easily predicted; in fact, it was predicted by me, and probably others. At least the U.S. suppliers had a much lower manufacturing cost than we did in Packard Warren. This was not the case with the Mexican suppliers, as compared to Packard Mexico. Virtually every Mexican company paid the minimum wages and benefits established by the Mexican government, so our Mexican suppliers had no employee compensation advantage over Packard's Mexican plants, and there is just no way they were going to be more productive than our plants were. After all, we had been in this business for decades and they were just getting started. They also had every other cost that we incurred in manufacturing a wiring harness: a plant, equipment and tooling, people to run the plant, utilities, and so on.

About the only cost advantage they could have over Packard's operations would be to build a cheaper plant that was not up to our specifications, but even this turned out not to be the case. They also had to create the same organizations and the same administration burden that we had, except for the services that we provided to them, which was an additional cost for us. So, the bottom line is that the Mexican suppliers had roughly the same cost profile that Packard's Mexican Operations experienced, but they also had to make a profit on every harness they sold back to us or they couldn't stay in business.

I once saw the numbers that showed that we were paying $2 more per standard hour of labor for wiring harnesses coming from our Mexican suppliers than what our cost was from our own plants. Not necessarily a great program.

When I was in working in South Korea, I once got a call from the vice president of one of the suppliers that I helped start up. He wanted to meet with me since he was in South Korea on business, and we had become good friends while we worked together in Mexico. What I found out was that they were no longer doing business with Packard, but that they now had several factories (when they were a supplier for Packard, they had none) and over 2000 hourly workers building products that were in direct competition with Packard. In fact, he was in South Korea working with a company who intended to give his company a large new contract. This was not hard to predict.

So, why in the world did Packard make the decision to set up a Mexican supplier program? All I can figure is that it was the decision of the head honcho who was making a classic, short-term decision based on bad data (maybe he erroneously thought we were going to save a buck or two as we had with U.S. suppliers), and no one wanted to tell him that it was a very unwise course of action, which would have been a real career-limiting act.

Mexico, Here We Come

In case you haven't already guessed, the second part of the new strategy, after the Mississippi strategy fell short, was the Mexico strategy. Speaking as a true blue American, it is unfortunate that this strategy was necessary. If the union had allowed the southern strategy to succeed, it would not have been necessary to foray into Mexico. Scores of additional plants would have been built in lower wage areas of the United States, providing great jobs for tens of thousands of Americans, all the while improving the infrastructure and the standard of living throughout the entire areas. However, that was not to be. The unions are even more shortsighted than management, and that is saying something.

I know that a lot of people were very critical of Packard when it first went to Mexico and built its first plant. How could a U.S. company do this to American workers? Well, the truth is that if it had not been done, Packard would have gone out of business long ago with the first wave of factories that were closed by GM, and everyone working in Warren, Ohio, and Clinton, Mississippi, would have been out of a job. The fact that Delphi Corporation held on as long as it did was in large part due to Packard, because it had been by far the most aggressive of the Delphi companies in moving into Mexico and other low wage rate countries. It is rather sad for me to think that, with all of the mistakes Packard made over the years, it was still at the head of the class within Delphi Automotive.

Black Friday

Nearly everyone has heard of Black Tuesday (or Black Thursday, the first big down day of the 1929 stock market crash). Well, in late 1977, an event occurred at Packard that I will refer to as Black Friday (at least I think it happened on Friday). You will remember the importance I placed on Packard's policy of not promoting operations personnel to positions of leadership without having prior Manufacturing and Industrial/Methods Engineering experience. This policy, while not perfect, had a very positive impact on ensuring that managers who were placed into positions of authority had some grasp on reality, as well as a good understanding of the business and the necessary relationships between the various operations departments. Something happened in late 1977 that necessitated a change in this policy, and this resulted in some very damaging, long-term consequences for Packard.

By way of background, it seemed to me that from the time I started with Packard every effort had been made by the company to be very fair and consistent in dealing with employees. In May of 1978, Packard's Management Council published the Packard Principles. Two of those principles are: (1) respect for the individual; it is important to recognize the dignity of every person, that they are unique and have the ability to contribute to the organization; and (2) trust in relationships; it is important to build and maintain mutual trust as it is the foundation for effective relationships. While these principles were captured in black and white in 1978, I believe they were practiced during my early tenure at Packard and, I am confident, well before.

I never saw any indication that there was any discrimination practiced at Packard against women or minorities (see sidebar).

Being a white male, some of you might be tempted to say, "Well, you wouldn't," but that would be very unfair. If discrimination is being practiced, any fair-minded individual will notice it, and I am fair minded.

If anything, what I observed is that being a woman or a minority gave an individual the tie-break advantage. Women and minorities were very well represented in all areas of administration and in some of the operations areas, especially where an engineering degree was not a strong desirable. There were certainly some women and minorities in the engineering areas, more every year; however, there were fewer than in the other areas, not because of discriminatory practices, but because of the shortage of available women and minorities coming out of college with engineering degrees.

What happened on Black Friday was that there was a meeting held in the Engineering Building, and every supervisory-level person and above who worked in the building was invited (i.e., mandated) to attend. In this meeting, the Operations director, who was the second in command at Packard, and who would later become the first managing director of the new Saturn Plant, told us that there would be some changes in personnel policy going forward, especially as to how new openings and promotions were to be filled. He explained that because we were part of GM, the world's largest company, we were under intense government scrutiny regarding adherence to EEOC guidelines. I don't think he used the word "quota," but it was very apparent what was happening. He first tried to explain the red block system. He indicated that if, per EEOC guidelines, there was an underrepresentation of women or minorities within a company at different levels, red blocks would be placed beside hiring or promotional openings, as appropriate. No white male could be considered until that red block was removed by first hiring or promoting a woman or a minority (minority women were especially valuable as they would offset two red blocks). However, according to the presenter, this was not a quota system. That just wouldn't be fair.

He went on to explain that Packard would never sacrifice its policy of putting the most qualified person into any new opening; it would just have to be done under the new red block guidelines. In my most politically incorrect manner, I asked the obvious question of what would happen if there were a conflict between Packard's policy of hiring or promoting the most qualified person versus this new red block system. You could have heard a pin drop. After a moment he gathered himself and stated, "There just won't be any conflicts." I'm sure everyone else in the room was just as satisfied with this answer as I was.

There were two areas where this new red block policy was particularly painful. The first was in engineering. When openings came up in engineering and there were red blocks required to be filled at that level, rarely were there qualified minorities or females to fill them. As you can imagine, the competition for qualified minorities and females with an engineering background was fierce. GM was not the only company under the gun. This meant that within Packard's engineering areas necessary positions went unfilled, sometimes for extended periods of time, until some other area filled the red block and opened the position in engineering to be filled by the best available candidate. But, I think that the most damaging consequence was what happened over time to Packard's top executive levels.

In the past, as I have explained, executive openings in the operations areas were virtually always filled by individuals who had proved themselves in engineering and manufacturing on the plant floor. In light of the new red block system, this policy was no longer plausible, because there were so few women and minorities with this background. From this point on, promotions into these positions were apparently made without consideration for real-world experience on the plant floor, regardless of whether the person being promoted was a minority, a female, or a white male. I don't think anyone was ever promoted into one of these positions of responsibility that was not smart and capable, but many were very inexperienced in comparison to earlier times. Many just did not know the wiring harness business very well, and, when that is the case, it is much easier to be swayed in your thinking by sweet sounding theory that may have no basis in reality.

This dramatic change in policy did not alter the quality of Packard management overnight. Promotions to these levels didn't occur every day, but, as I look back and analyze Packard's past, this was certainly a turning point. It seemed that every year, from this point on, the quality of the decisions coming out of Packard's executive levels regarding wiring harness manufacturing grew weaker and weaker; decisions that would not have been made a few years earlier. (Many examples of this will be highlighted later in this book.) It seemed that from this point on Packard's management had less and less confidence in its own abilities to make good decisions, even though Packard had been in the wiring harness business for over half a century. Instead, Packard executives started relying more and more on theories of Lean experts who had never before set foot in a wiring harness factory.

Something Had to Change for GM

In the early 1990s, GM knew something drastic needed to be done with its vertically integrated suppliers if it was going to survive. It just could not continue to pay for components that were built with high-cost union labor, when its competitors were not. A decision was made to sell some of the vertically integrated companies, close others, and consolidate those remaining into a new division called the Automotive Components Group. This division was renamed Delphi Automotive Systems the next year.

And then the fateful day arrived in 1999 when GM made the decision to spin off Delphi Automotive Systems as an independent company. I remember many Packard employees expressing a lot of excitement and enthusiasm that day because we would no longer be under the shackles of GM and we were now free to compete in the entire automobile market; it would no longer be so difficult to win business from GM competitors.

I was much more reserved and the old adage came to mind: "You had better be careful what you wish for, because you just might get it." I had had the opportunity to see a lot more manufacturing operations than most of my co-workers, including the facilities of competitors, and I knew our competitive position was nowhere near what some thought, even with our low-cost operations in Mexico. And, even more troubling, Packard was at the head of the Delphi pack based on all observations, financial results, and anecdotal evidence. None of the other Delphi companies had anywhere near the presence that Packard did in low-cost manufacturing locations.

Packard Goes International in a Big Way

Starting with the Mexican Operations in 1978, Packard accelerated its strategy of moving more and more of its business to low-cost countries through wholly owned operations and joint ventures as well as growing its business around the world through acquisitions and joint ventures. It is hard to fault this strategy. Packard intended to remain a viable company and, to do so, it needed to participate in the world economy. I have been personally involved in the Mexican Operations, and I know that Packard had no choice but to pursue this strategy after the plan of building facilities in low-cost U.S. locations bombed out.

I spent two and a half years in Packard's Portuguese facilities in the mid-1980s, which were part of Packard's European Operations, headquartered in Germany. These were purchased by Packard in the early 1980s to get a foothold in the European market. Facilities in Portugal and Ireland were the low-cost producers for Germany at that time, which suffered from some of the same noncompetitive problems that Packard faced in the United States. I then spent four years in South Korea with Shin Sung Packard in the late 1980s, a joint venture that Packard established to service the South Korean automotive market. Packard now has facilities in China, Malaysia, Romania, Hungary, Turkey, Morocco, Austria, Australia, Brazil, and several other countries.

I do believe that this was a strategy that had to be pursued by Packard. It didn't really have a lot of choice, and from what I understand, most of these facilities are doing fairly well and making money. Packard wisely understood that if it was to have a future, it was not going to be in U.S. manufacturing; it just was not going to be possible with the union situation when competitors did not have that "obstacle." Packard had no choice but to get footholds in markets around the world and establish manufacturing facilities in low-cost countries. I think it has done a pretty good job of doing this.

What it has not done a good job of doing is recognizing that it is a manufacturing company first and foremost, and, if operations are not set up in the right way with the proper systems and procedures, it is just a matter of time before they, too, have severe problems. In the next chapter, I will talk about some of the major mistakes that Packard made, some of which have been transplanted to these other operations, which will one day create big problems. My hope is that other U.S. companies can learn from these mistakes.

We Are Not Ready for This

It didn't take long for Delphi and Packard to realize that the sledding was going to be pretty tough, now that they no longer had a virtual guarantee of GM business. And, the GM business they did have didn't look like it was going to be nearly as profitable, since GM was demanding significant price reductions every year.

In truth, GM was largely responsible for the Delphi companies being non-competitive. GM was responsible for Delphi Companies being saddled with old, decaying plants in very high-cost manufacturing areas with a union workforce that, along with their leaders, just didn't get it. Looking back, it should not have taken a crystal ball to predict what would eventually happen (and not too many years down the road, at that).

The situation with Delphi is a little like a person growing up in a home where he is coddled his entire adolescent life, given little opportunity for real-life experiences, given little education and training, and then shoved out the door at age 18 with a pat on the back and a good luck wish.

Aside from the obvious, there were many other areas in which Packard (and other Delphi companies) were not ready to compete. One example, with which I am very familiar, was Packard's system for costing and pricing wiring harnesses, which was Packard's main business. The system that Packard had used throughout the years was one in which the labor content for each harness was estimated using standard data, which was then adjusted based on a factor derived from the known labor content of a similar harness, as compared to the standard data estimate for that similar harness. So far, so good. This labor content was then multiplied by the average cost per standard hour of labor, inclusive of all burden, and then increased by the material cost. The cost was then marked up by a suitable profit percentage to establish the price.

First of all, Toyota has demonstrated that the real equation should be: Price (as determined by the market) − Cost = Profit, not Cost + Profit = Price, as had been Packard's practice. However, this was not the big problem. The Packard system made no allowance for volume or for part number proliferation in its pricing system. This had been a concern of mine as early as the 1970s when Industrial Engineering was doing the estimating as well as doing the preplanning (the joke was that it was easier to create improvement with the pencil than it was through good engineering in the preplan process).

Our system was motivating exactly the wrong kind of behavior from our customers. Because there was no penalty associated with increased part numbers, it was cheaper for GM to add a part number if a vehicle option required one more or one less circuit than it would be for them to use the same part number for both options and have one unused circuit for one of the part numbers. This resulted in some harness packages (e.g., Cadillac DeVille Instrument Panel Harness) that had 30 or more part numbers, each of which greatly complicated the manufacturing process and drove up the cost.

When I returned from South Korea in 1990, I quickly found out that the system had not been improved, so I wrote the following letter to the appropriate Packard executives:

> Packard's current estimating and costing system does not take into account the negative impact of low volume and part number proliferation. As a result, our customers are encouraged to increase part numbers in order to avoid unused content, and thereby save money (even though our cost is increased).
>
> While we have not quantified the precise cost to Packard resulting from part number proliferation, we know intuitively that it is considerable. Our current system misleads our customers and encourages a violation of one of the most critical Design for Manufacturability (DFM) principles—reduce part numbers.
>
> While Packard's total sales revenue is probably in the ball park, we are rewarding our low volume, part number proliferating customers by subsidizing them, at the expense of our good customers (i.e., high volume and low part numbers). I believe that this inequity must be corrected ASAP, understanding that it will take some time to develop and transition into a revised system and communicate it properly to our customers.

I attached two matrices to the letter that contained my best SWAG (scientific wild-assed guess) as to the true penalties we were incurring due to low-volume packages and to part number proliferation. With a lot of arm twisting, Packard finally made a half-hearted effort to capture the impact of part number proliferation (except that it was done the wrong way and Packard didn't get all of the positive results it should have), but Packard did nothing to consider the impact of volume. As one Packard executive told me, "We're just not going to go into

Cadillac and tell them their prices will increase." I told him that was fine, but we probably shouldn't tell Chevrolet that they were subsidizing Cadillac, either.

Packard could get away with that type of costing/pricing system when virtually all of their business was guaranteed GM business. Because everyone was GM, it was just a matter of which pocket the money went into. But, this was certainly not the case when Packard was spun off, along with the rest of the Delphi group. Packard would have never developed a costing/pricing system like this if it had been an independent company from the beginning. There is just no way a company can survive by under-pricing unattractive business (low volume, high part numbers), which they will undoubtedly win and then lose money on, while, at the same time, over-pricing attractive business (high volume, low part numbers), which will not be won. This is a recipe for failure.

We Are in Real Trouble Here

Packard and the rest of Delphi realized they were in real trouble with their U.S. business very quickly after the spin-off. Early retirements had been offered to certain Delphi salary employees (generally, those aged 58 and above with 30 or more years of service) at the time of the spin-off, and they were allowed to retire as GM employees. It looks like those that took the deal made a good decision. By early 2001, early retirements were being offered to those 55 and older with 30 years, and in June of 2001, the age dropped to 52 with 30 years of service. I decided to get on board at this point, as it became more and more clear to me that Packard was on the wrong course in so many ways and there was nothing I was going to be able to do to change it.

Over the following years, Packard had several other early retirement pushes, each one providing stronger "encouragement" than the last. Based on some of the letters and blogs I have read, it appears that some Packard salary employees were virtually pushed out the door, even though they had no desire or intent to retire at that time.

In October of 2005, Delphi filed for Chapter 11 protection, and it was in the reorganization process for almost four years before emerging. However, it was only for their North American Operations. Their profitable International Operations were somehow excluded from the bankruptcy. I didn't even know this could be done. This shouldn't have come as too big a surprise, what with all I knew about Delphi and its problems, but to some degree it was. I remember my dad telling me when I took a job with Packard (which was part of the GM family) that I would never have to worry about a job—GM would be around forever. Even though all logic and common sense indicated that there was no way Delphi could avoid bankruptcy, especially when they brought in an outsider as the new president of Delphi who had taken other companies through Chapter 11, I was still somewhat surprised when it happened, I guess because my father's words were still floating around in the back of my head.

When I retired from Delphi, I signed a bunch of papers, like everyone else. One of the things that Delphi was very careful to do was put wording in these papers to the effect that, with the exception of our pensions, which were fully vested and into which we had been contributing our entire working lives, Packard reserved the right to amend, modify, suspend, or terminate any of its benefit plans or programs, which retirees had earned during their working lives and were promised in retirement. (While Packard was still part of GM, these devastating words were rarely included on the benefits statements we received, which instead promised that the benefits would be for life.) The implication for all of us, who retired early, in an effort to try to help the company with its current financial problems, was that this was only legalese and nothing to worry about. Packard had no intention of eliminating these benefits, or so we were told and so we thought.

In 2007, while still in Chapter 11, Packard made the decision to terminate any benefits to salary retirees who had reached the age of 65 and were eligible for Medicare (this included all healthcare coverage, dental and vision coverage, and life insurance). Even though we were told by Packard that our salaries were lower during our careers to offset the cost of lifetime benefits (which was further substantiated when Packard reduced retirement benefits for those hired after 1993 and gave them a salary increase to offset it), I knew that Packard was in big trouble and needed to find some ways to cut costs. I'm not aware of any organized effort to challenge this action. I know I didn't. I only wondered why it was only salary retirees that were affected.

In late 2008, Packard and Delphi had a final big push to get as many salary employees as possible to "voluntarily" retire before making the big announcement. In February of 2009, Delphi announced that effective April 1, all salary retirees' benefits would be terminated—forever. You can imagine the shockwaves that this sent through the retiree community. There was no warning. Many of the individuals agreeing to an early retirement would not have made that decision (if given the option) if they had known that they would lose all benefits, especially those who were pushed out the door in late 2008. This was a move that has devastated many of the retirees, who weren't really ready to retire, and now find it almost impossible to make ends meet or to find another job.

This action on the part of Delphi did draw a response from the salary retirees. A Delphi Salary Retirees Association was formed, which is challenging Delphi's decision in court. On March 11, Judge Drain ruled in Delphi's favor and all salary retirees' benefits were terminated effective April 1. The decision is still being appealed, but it looks pretty certain that those benefits earned through a lifetime of hard work and dedication are gone forever. The sad thing is that I am confident salary retirees would have been willing to make very significant sacrifices to help Delphi get back on solid footing if they had been approached, especially if commensurate sacrifices were to be made by the union. However, the UAW emerged virtually unscathed and the IUE and some smaller unions got a haircut. What kind of a message does that send? I am reminded of two of the Packard Principles that were detailed earlier, namely Respect for the Individual and Trust in Relationships. What can one say?

As if that weren't bad enough, part of the Delphi plan to emerge from bankruptcy was to turn over all its pension liabilities and assets to the Pension Benefit Guarantee Trust Corporation (PBGC), which is a government agency created in 1974 as part of the Employee Retirement Income Security Act. This agency collects premiums as set by Congress from companies having defined pension benefit plans, and, in the event of a participating company's bankruptcy or default, it takes over the company's pension assets and obligations. This may sound pretty good, but there are a few problems. For starters, the PBGC is currently operating at a deficit of over $33 billion. Secondly, there are defined limits that the PBGC will pay as set by Congress based on a recipient's age and other factors, which usually preclude any early retirement incentives or bridges until Social Security kicks in. Delphi's salary retirees nearing Social Security age were not expecting to see big reductions, but most of us have received big shocks (reductions of 30% or more appear to be the norm), while younger retirees, most of whom were unceremoniously pushed out the door, are seeing drastic reductions, some as much as 70%. This, coupled with the loss of all benefit, will push many retirees below the poverty level, and jobs for those 50 and older are not all that easy to come by these days.

This is a pretty bitter pill to swallow, but it gets worse. GM decided that it had an obligation to Delphi's unionized hourly retirees that it did not feel it had to Delphi's salary retirees, even though most of these salary retirees had worked about three decades of their careers as GM employees before being spun off. GM made the decision (undoubtedly with the full knowledge and approval and possible coercion of its primary owner, the U.S. government) to "top-up" the pensions of all Delphi's UAW retirees (obviously with some of the money it got from the government, i.e., your money).

When the IUE and the other smaller unions squawked, GM decided to top them up as well. However, GM could muster no such sympathy or feelings of obligation for the salary retirees. The newly formed Delphi Salary Retirees Group has appealed this decision and it is still in the courts while waiting for information and documents from both GM and the Obama administration that would cast light on how this decision was made and justified.

I have provided this information not necessarily to solicit sympathy for Delphi retirees, but so that the reader will have a better understanding of just what kind of company this is, how it operates, and how decisions are made. I have written a lot about the shortsightedness of GM and the GM subsidiary companies, including Packard, and I'm wondering if you can see the same pattern emerging, remembering especially that Delphi is not only a manufacturing company, but also a technology company. What is the likely impact on Delphi and its ability to recruit, hire, and then keep talented people in the future? What is the likely attitude of the majority of salary employees and their commitment to Packard/Delphi? I have already heard rumblings of the need for a salary union, which would be a disaster.

Delphi and Packard cannot get out of the mess they are in on the backs of their salary retirees; it is going to take major concessions on the part of the union as well as an end to Delphi and Packard making so many bad decisions. What they have succeeded in doing now is making enemies of 15,000 salary retirees, who were former ambassadors, along with everyone else that these retirees can influence.

Industrial Engineering (IE) Training Program

Earlier in this chapter, I talked about the IE Training Program, which every engineer new to the IE Department received prior to being given an assignment. When I returned from South Korea in 1990, I discovered that no training of this type was being provided to new IEs in the Mexican Operations, even though it had been done at least through 1983 when I first went overseas. One of the first things I did upon returning to the Mexican Operations was to develop a training manual and try to re-institutionalize the training program, but with limited success. Over time I wrote, organized, and compiled a several hundred-page IE Manual, complete with the training program. However, the IE training program was never again institutionalized at Packard to anywhere near the level it had once been. The new group in charge just didn't understand the importance of this critical training.

The following (including Figure 2.1.) was included in the Introduction section of the manual that spelled out the contents of the manual and the training program outline.

The Industrial Engineering Training Program

Training can either be an invaluable activity that will provide great benefit and skill enhancement to the participants (including the trainers), or it can be a total waste of time and money, or worse yet, it can mislead participants into developing and carrying out nonproductive/nonvalue-added activities and initiatives. The attached training outline (which closely follows the contents of the Industrial Engineering Manual) has been proved through many years of experience to be the former case and not the latter.

Regardless of how good an education an individual receives or how much experience an individual has in other companies, a certain amount of company-specific, hands-on training is necessary if that individual is to be productive in a reasonable amount of time. And, the earlier an individual receives this training the better, before the new hire gets totally lost and confused and potentially makes errors that could jeopardize customer satisfaction or critical company performance measures.

This is especially true in Industrial Engineering. In fact, it is true for anyone entering the IE department, whether directly from college (even if they have an IE degree), from another company (even if they were IEs in that company), or if they are transferred into IE from another department within our company (*regardless of which department*). As IEs, the quality of our work has a tremendous impact, not only on customer satisfaction issues (quality, delivery, responsiveness, etc.), but also very much so on the company's critical performance issues (productivity, cost, quality, health and safety, etc.).

The following three-week training program is patterned after an IE training program, which was utilized successfully in the past to properly train and orient each new IE prior to him/her being given any IE assignment. The result was that virtually every IE, even those fresh out of school, was fairly productive from the day he/she received the first assignment, and he/she continued to become more and more effective with practical experience. It was also rare that critical mistakes were made that had major impacts on customers' satisfaction or key performance issues; mistakes tended to be of the less serious "learn from your experience" variety.

Trainee(s)		Date	
	IE Training Program	Page	1 of 2
		Scheduled by	

Topic	Training Location	Instructor	Training Duration		Day	Date	Starting Time
			Full	Condensed			
1. Problem Solving Techniques 7-Step Engineering Approach Scientific Method 5-Whys Process 5-Phase & Shanin Technique 8D Process			3.0 hrs		1		8am
2. Basic IE Tools Ergonomics Ineffective Worker Movements NECEB, Peg and Washer Act Breakdown			5.0 hrs				11am
3. Work Measurement Standard Time Data Cycle Check			2.0 hrs		2		8am
Check Study			4.0 hrs				10am
Master Study Conveyor Master Study Stop and Reject Study			2.0 hrs				3pm
Delay Study			8.0 hrs		3		8am
4. Economics Value Analysis Economation Concept Economic Evaluations Cost and Investment Reduction			4.0 hrs		4		8am
5. Lean Concepts 4 Key Questions Sub-Optimization Value Stream Analysis True TPCT Reduction Six Sigma			4.0 hrs				1pm
Lean Workshop			4.0 hrs		5		8am
7. Engineering Organization Key Responsibilities Interfacing with other areas Major focus, goals, objectives History, lessons learned			4.0 hrs				1pm
6. Quality Systems & Procedures Quality Philosophy Quality Measurements			2.0 hrs		6		8am
8. Manufacturing Support Manufacturing Organization Plant Priority Meeting Forward Planning On-the-floor support			2.0 hrs				10pm
9. Lead Time Reduction Program Trackers			4.0 hrs				1pm

Figure 2.1 Form.

Trainee(s) _____

_____

_____

IE Training Program

Date _____
Page   2 of 2
Scheduled by

Topic	Training Location	Instructor	Training Duration		Day	Date	Starting Time
			Full	Condensed			
10. Preplan Process			2.0 hrs		7		8am
Preplan Schedule							
Harness Design Analysis			2.0 hrs				10am
DFM, IECR							
Pegboard Review & Build M.B.			2.0 hrs				1pm
DFM/DFA, Order Repair Tools							
Build Sequence			2.0 hrs				3pm
CTS, Pre and Post Operations							
Work Content			8.0 hrs		8		8am
Production Plan & WC Meeting			4.0 hrs		9		8am
SB Data							
Paper Breakdown							
Rough Breakdown & RB Mtg.			10.0 hrs				1pm
Station Layouts							
Final Breakdown & FB Mtg.			10.0 hrs		10		3pm
Conveyor Data, Packaging Data							
Visual Aids							
Area Layout and Flow			2.0 hrs		12		8am
Line Layout, Rack Order							
PFMEA & Flow Diagram			2.0 hrs				10am
Master Board Release			4.0 hrs				1pm
WLL & MB Release Sheet							
System Release & Start-up			4.0 hrs		13		8am
Supervisor's Notebook							
Line Start-up							
LS Chart & Graph, Follow-up							
11. Productivity System			8.0 hrs				1pm
Philosophy							
Labor Standards							
Direct Labor Bible (DLB)							
Production Efficiency							
Process & Operator Efficiencies							
12. Labor Estimates			4.0 hrs		14		1pm
Standard Data							
Budget DLB							
Budget Improvement							
Plant Improvement							
13. Control Sheets			4.0 hrs		15		8am
Controls 1, 2, 3, 4, 5, & 6							
14. Engineering Change Control			4.0 hrs				1pm
Eng. Change Control Meetings							
Trackers							

Figure 2.1 (*Continued*)

To ensure the most effective training possible, adhere to the following basics:

- Every new salary employee assigned to Industrial Engineering (regardless from where they come) should go through the complete three-week training program **prior** to being given any assignment.
- This should apply to new salaried IEs at any level, unless they have prior IE experience within our company. Even then, an abbreviated refresher course should be considered.
- Don't wait for a full classroom of new IEs to conduct the training program. Conduct it for one person, if necessary. In fact, it is normally more effective when conducted for one person.
- The training sessions, which are the responsibility of IE, should be conducted by practicing IEs, not by designated trainers.
- Every Methods and Industrial Engineer should be able to teach each and every section of the IE Training Program. If they are not competent to do so, they need to get competent.
- This will more fully engrain the concepts into the minds of the trainers, as everyone knows that the teacher learns more than the student in any training exercise.
- This also will ensure that the new engineers are taught realistic and practical concepts, not theory from someone who hasn't or isn't doing it.
- The three-week schedule should be developed and given to the trainee on the first day of the training program, complete with subject matter, timing, location, and name of trainers (see Figure 2.1).

Note 1: The IE Training Program will be of great benefit to any employee regardless of his or her current assignment. This training also should be made available to other areas and conducted as time permits.

Note 2: A condensed training program can be developed, scheduled, and provided to individuals within IE with specific weaknesses or needs, or to provide a review. A condensed training schedule also can be used to provide individuals outside of IE with a general understanding of IE systems, procedures, concepts, and tools.

Chapter 3

How Does a Company with So Many Smart People Do So Many Dumb Things?

When I was growing up, and very prone to the foibles of youth, my father would often ask me why such a smart boy as I did so many dumb things. It was certainly a fair question, as I would frequently do things to put life, limb, or property in jeopardy—time and again.

Having spent 30 years of my career working with Delphi Packard, I could ask the same question. How could a company with so many smart people make so many dumb decisions? There is no question that there were a lot of very talented and very smart people working within Packard. Being part of GM, and being a successful manufacturing company with very impressive engineering credentials in component design and testing, product design, and manufacturing system design, Packard was able to attract some of the best talent from some of the best schools in the country. Although, based on what has recently transpired, I am not so sure this will ever again be the case. However, throughout my 30-year career with Packard, it seemed that an inordinate number of really dumb decisions were made.

One might ask why dredge up the mistakes of the past? Some of these things happened a long time ago. Is it relevant to even talk about them today? To both questions, I would give a resounding "yes." What I have learned in life is that the best way to avoid mistakes tomorrow is to learn from the mistakes of yesterday. We have often heard, "If we don't learn from the mistakes of history, we are bound to repeat them." I am confident that the kinds of problems Packard faced (and the kinds of mistakes that Packard made) are similar to ones being faced by companies every day, and problems that companies will continue to face in this very competitive world. To paraphrase something I have heard: "It is a smart person who learns from his mistakes, but it is a wise person who learns from the mistakes of others." I am trying to help those of you reading this book to become wiser, not just smarter.

Hence, why did Packard make so many boneheaded decisions? I think some of it goes back to what John DeLorean said, "Many executives who rise to the top (within GM) are not the best managers, but [are] the men [and women] most skillful in flattering their bosses. A system which puts emphasis on form, style, and unwavering support for the decisions of the boss almost always loses it perspective about an executive's business competence." But, a lot of it was due to what John McElroy wrote, "American companies don't seem to have the discipline or attention to detail to grind out consistent annual improvements … and ultimately succumb to distractions when management gets bored with consistency and starts dreaming of breakthrough programs that will leapfrog the competition."

However, I think there are two other major factors that are every bit as important as these. The first is that, due to the tremendous competition that existed within Packard to reach the highest levels of the company, there was a belief (based on history and observation) on the part of many career-minded individuals that the way to get ahead was to do something very spectacular and different than the norm, something that would get noticed, although, as often turned out, not something that was necessarily in the best interests of the company. Although occasionally an individual who took this route got left holding the bag when things didn't pan out, frequently what happened was that these individuals were rewarded for their initiative and innovation and were whisked away to new assignments, while others were left behind to clean up their messes when things fell apart.

The second is that from the day I started working at Packard, it was apparent that something drastic was going to have to be done, and in the not too distant future, if Warren, Ohio, was going to remain a viable manufacturing location. Many of the individuals in upper management at Packard at that time had very deep roots in northeast Ohio, and they did not want to see the Warren Operations go down the tubes. There was undoubtedly a fair amount of fear in the hearts of many of these managers who knew something spectacular had to be done to save Warren as a manufacturing base. Fear can be a great motivator, but, if not properly harnessed, it can spawn some very poor decisions and bad outcomes.

In this chapter, I am going to focus on some very poor (sometimes horrible) decisions that Packard made prior to the introduction of Lean Manufacturing concepts or Toyota Production System (TPS) concepts within Packard (in the mid-to-late 1980s), or to bad decisions that were made after that time that really didn't involve Lean Thinking. In the next chapter, we will review some of the nitwit things that were done in the name of Lean or the Packard Production System (patterned heavily after the TPS). The hope is that, by seeing some of these absurdities in black and white, others can avoid similar thinking and similar results.

Automation Can Save Us

In the late 1960s and early 1970s, Packard management knew they were going to have to do something about the high cost of union labor if Packard Warren was to remain viable as a wiring harness manufacturing site. The vast majority of Packard's some 13,000 hourly employees were unionized by the IUE (International Union of Electrical Workers), which marched lockstep with the UAW (United Automobile Workers), and had virtually identical benefit packages, wages, and mindsets. GM and Packard had had no success in scaling back union wages, benefits, and restrictive work rules. In fact, just the opposite was true, as each new contract continued to get more and more noncompetitive. It seemed that the only way to solve the problem, because it was obvious the union was never going to "get it," and GM wasn't going to get the gumption to stand up to the union, was to find a way to significantly reduce the amount of labor required to build a wiring harness.

With the excellent work that was being done in the Methods Lab, virtually all of the potential labor improvements had been achieved on final assembly build processes going into production. There just wasn't much blood left to squeeze out of that turnip. There were lots of opportunities in the cutting departments, which were very inefficient, but that didn't involve that many people relatively speaking (less than 10% of all hourly workers engaged in wiring harness production were employed cutting wire), and the union was not going to allow the changes that needed to be made, anyway. The lead preparation areas were much better organized and quite efficient (meaning that the standard hours generated in these areas based on fair standards were very close to the actual hours of labor used), and they also accounted for only about 25% of the hourly workers used in wiring harness manufacturing.

Upper management came to the conclusion that the only way to make a significant dent in the amount of labor required to build wiring harnesses was to automate. In the right environment, and, most especially, with the right products, automation can be very economical and provide some tremendous savings. However, there are some products and some situations where automation is just not the right answer. Unfortunately for Packard, the final assembly of a typical wiring harness just does not lend itself to automation, but this fact was lost on upper management.

Sometime in the late 1960s, three or four automated production lines were installed in one of the plants for a very nontypical wiring harness type. (They were really semiautomated because there were some operators assigned to the production lines to do some things that could not be done automatically.) The harness type was called Front to Rear, and, as the name implies, it was the harness that mated with the Instrument Panel Harness in the front of the car and carried power to the rear of the car, where it mated with the Rear Body Harness. This was an extremely simple and stable harness type, which was comprised of a five- or six-conductor solid core ripcord (ripcord is two or more wires that have been fused together, like the two-wire ripcord that runs from your table lamp to the outlet), covered with plastic tubing, generally having one simple breakout, with the terminated wires plugged into simple connectors on each end. The package was very high volume because virtually every GM car contained one, and there were very few part numbers because common platforms could use the same part numbers. The solid core wires also made this harness type ideal for automation (if, in fact, any are) in that solid core wires are much more rigid than stranded wire, so it was much easier to cut and terminate automatically and then plug automatically. (On a nonautomated line, wire is cut and terminated on wire cutting machines under the control of one operator and then brought to the line for manual assembly.)

In short, this was the ideal harness type for automation (if, in fact, that type existed). On my initial tours through Packard, I was shown these lines and, I must say, they were impressive for someone fresh out of college. But, even then, I wondered if they were justified. There was a lot of expensive tooling and equipment, there was a lot of floor space tied up, and, while there were only a couple of workers assigned to each line, there was a cadre of maintenance workers in the area. It also seemed that at least one of the lines was down every time I walked through the area.

Were these lines justified? Did they save money for Packard? Well, it depends on who you ask. These lines were generally touted as being a very good investment by Packard executives, but, if you talked to most of the engineers and manufacturing supervisors who were very familiar with the area, they presented a different picture. Their position was that the one or two direct operators eliminated by the automated (maybe better called "mechanized") lines were more than offset by the increased investment, floor space, maintenance, and other overhead. While I never did my own economic analysis, I would tend to believe the second group rather than the executives who touted the system. There is an old saying that can be very appropriately used here and it was often quoted by engineers at Packard: "Figures don't lie, but liars can figure." In other words, people can often prove about anything they want to prove by manipulating the data and being very selective in what is considered and what is not considered in the evaluation (our government has honed this skill to a science).

In my mind, even if these mechanized lines were not cost effective, the decision to try to automate this product was justified. Packard had to learn what could and couldn't be done with wiring harness automation. If this product could not be successfully automated, then none could. My understanding is that the first line was set up and run in the Process Lab and, when it appeared that the bugs had been worked out, they made the decision to install all of the required lines in the plant. My qualm is that we didn't need to install all three or four lines at first; at least until a very clear determination had been made as to whether or not the lines were justifiable. I'm not confident that that determination ever was made in a nonbiased manner.

That Worked So Well, Let's Try Something a Little Tougher

The next foray into automation/mechanization occurred in the early 1970s in the assembly plant in which I had recently been assigned as a manufacturing supervisor, so I had the opportunity to see the results of this catastrophe first hand and often. I am just glad it wasn't in my department.

Packard management had determined the results of the Front to Rear solid core lines to be a success (even though it was far from a universal opinion), and determined that it was time to take the next step in semiautomation. I didn't then, and I don't now, have a problem with Packard's decision to make additional attempts at automation (or semiautomation, mechanization, or call it what you will), but I have a gigantic problem with the package that was selected and the way the program was implemented. One of the process rules of Toyota is: Use only reliable, thoroughly tested technology that serves your people and processes. Again, this is just good common sense; unfortunately, again in this case, Packard just did not use good common sense.

The package that was selected was the Chevy Impala Forward Lamp Harness. The harness itself is not overly complex, containing only 25 or so circuits, but it is a fairly long and gangly harness containing two splices running the entire length. It would have been much better to select a simpler harness like one of the accessory harnesses (such as an air conditioner harness or a power window or door harness) or even a small engine harness. The biggest mistakes made, however, were in selecting a package that had a volume of somewhere around 3000 harnesses per day and in not having a good backup plan in case problems arose. (Hard to believe that the full-size Chevrolet sold at that volume level at one time when it only sells at a few hundred per day now.)

There was no cost-saving technology available then (nor will there ever be) for wiring harnesses of current design that can accomplish the automatic stringing of wires onto a build fixture, the plugging of those wires to connectors, and the applying of the covering, along with the application of any necessary clips, clamps, etc. All that could then be done automatically was to cut and terminate the wires, plug the terminated ends of the wires into plastic connectors, and then convey that subassembly to an operator, who would then place it onto the build fixture and do additional final assembly work as his/her station time allowed. Apparently, some testing had been done in the Process Lab and the subassemblies had been preplanned into the build method in the Methods Lab, but the process was far from debugged.

Toyota would have never gone into production with such a poorly designed and prepared process, but we did, and, boy, did I ever feel sorry for the supervisor of that Final Assembly Department. Because of the excellent preplan jobs that were done in establishing final assembly build methods, most model change startups at Packard went very smoothly, even when starting up new vehicle programs or vehicle programs that had undergone major change. However, this one was a disaster. The on-line equipment was down more than it was up. The only way to keep the lines running was to have the subassemblies made in another area and then brought to the line on portable racks and somehow squeezed into the line, so that the operator could take them from the racks and complete her (at that time, most conveyor operators were women) station. And, of course, this nonsense did not improve the prospects for good quality or good operator morale.

Even if the automated portion of the process had 100% uptime, there is no way that the meager reduction in direct labor could have been offset by all of the investment, overhead, and maintenance requirements. Anyone with any knowledge of manufacturing processes could have determined this without putting a single number on a piece of paper. But, with the way the process was running, it was clear that Packard had put a system in place that had taken a very profitable product and turned it into a big loser. Packard could certainly absorb this kind of loss (at that time), but it was still a terrible decision to proceed in a manufacturing environment with such a poorly designed and non-debugged process.

What should have happened? Like Simon Cowell tells a lot of singers on *American Idol*, "You picked the wrong song," or, in this case, the wrong product. It was the wrong product from a design standpoint and from a volume standpoint. The Packard thinking must have been, "Hey, we may lose a ton of money on every harness we make, but we will make it up on volume." This was another running joke within the engineering community.

Most importantly, this process should have never seen the light of day in a manufacturing environment until it had been fully debugged and successfully run in the Process Lab, or at least some quarantined area (although, using factory workers to run a pilot program is the right answer). Packard management never seemed to grasp the damage done to the psyche of production operators, not to mention engineers and manufacturing supervisors tasked with making things run, when really stupid things were done in the plant that didn't work. It certainly leads to reduced enthusiasm "to bust backsides" to be productive when the systems put in place by management make that an impossible task.

I don't fault Packard management for making the decision to spend money on trying to develop new technology, even when it was obvious that it was not cost justifiable in current form. It's not always cheap to conduct tests, experiments, studies, and pilots in an effort to gain new knowledge, which can potentially give you an edge in the market place. However, it must be done in a responsible way and without jeopardizing the customer, the attitude of our employees, or the reputation of our company.

That Lead Prep Startup Surely Didn't Work Out Like We Had Planned

In the next section, I'm going to talk about a decision that was made on behalf of the Mississippi Operations (in reality, it was harshly forced on them), but, in this section, I am going to talk briefly about a decision the Mississippi Operations made where it shot itself in the foot; a decision which would leave one to wonder: "Why on Earth did they do it?"

For decades, standard technology in the wiring harness business had been the use of semiautomatic wire-cutting machines that would, after the appropriate setup, cut and strip wires (leads) to the appropriate length, and then automatically apply terminals to the ends of the leads. Semiautomatic wire cutting and terminating had become a very reliable process over the years and was the standard for the industry.

That's not to say that no problems existed with semiautomatic cutting and terminating, but significant problems also existed with the technology with which they chose to replace it. To give an analogy you might understand, it was a little like giving up your TV remote control in favor of the "if you want the channel changed, get up and change it" technology we used to have, because sometimes batteries go dead or there is occasionally some other malfunction. What they decided to do was use the semiautomatic cutters to cut and strip the wires only and then send the wires over to the appropriate Lead Prep Department to have the terminals applied at manual sit-down presses.

In theory, there were some potential benefits in making this decision. For one, by eliminating the need to install dies on the cutters, the setup time when going from one lead code to another would be reduced on the cutters. It might reduce the setup time by half of the current 8 to 15 minutes, depending on the degree of change from the prior lead code. This had the potential to reduce the need for a cutting machine or two (which at $50,000 or so a pop wasn't insignificant). However, when you consider that if everything worked out according to plan and the manual terminating was done at 100% efficiency versus current standards, it still meant that the labor to cut and terminate lead codes would more than double, and this would not be a one time thing; it would go on forever. And, let's not forget that the additional operators would require additional presses and dies and floor space to do their work, not to mention the fact that we put another step in the process flow, complete with an additional inventory location.

There also were some who felt that quality improvement might result. On rare occasions, a crimp over insulation would be encountered that came from a cutting machine. This meant that the insulation was not properly stripped from the wire and the terminal was crimped over the insulation. Most of the time, these were found during the harness electrical test, because they would not make an electrical contact, but very rarely a terminal would cut through the insulation and make intermittent contact. This could lead to a harness passing an electrical test, but showing up as an electrical problem in the assembly plant. However, the same problem also could exist on a manual press, to a greater or lesser extent depending on the experience level and contentiousness of the operator. This problem was later solved altogether through crimp-monitoring technology.

I don't know who pushed the proposal to scrap the semiautomated cutting and terminating technology in favor of technology that hadn't been in vogue for decades, but, ultimately, the decision was made all the way up the organization. This was just another example of an unbelievably poor decision being made within the company. I can understand an engineer proposing this plan, so long as he/she didn't bother to do a good cost–benefit analysis, but I cannot imagine that anyone who had experience in a Cutting or Lead Prep Department would have been part of it.

So, what ultimately happened? One of the monumental startup projection fallacies was in the estimate of the time it would take a new operator to become efficient on a sit-down terminal application press. Anyone with experience knew that it would take a good eight weeks for an operator to become highly skilled on a sit-down terminal press. Sit-down press jobs were coveted at Packard for direct operators because, as the name implies, the operators would sit down while doing their job. Whereas, virtually all of the other direct operators had stand-up jobs, e.g., on the conveyor line, on off-line build stations, or even on the cutting machines. These jobs were obtained by very senior operators and were rarely given up unless someone retired or was incapacitated.

The estimate used on how long it would take to become efficient on manual presses for the Mississippi Operation was only a fraction of the real time. As a result, the startup was a colossal mess. Lead codes were not getting to the lines in time and lines were shutting down all over the place for lack of material. When I was a foreman a few years earlier, I, and virtually all other foremen, considered it almost a capital offense to have a line down for lack of material. You can imagine the image this presented to the new, inexperienced operators trying to learn their new conveyor jobs in a brand new plant.

Again, this was easily predictable, and several people did predict the inevitable result, including me. The sad thing is that again there was no way that this decision could have ever been justified, even if the new press operators ran at 100% efficiency from day 1. Obviously, whoever made the ultimate decision was listening to the wrong people. There may have been some very unpleasant fallout for some individuals, but I never heard about it if there was, which leads me to believe the decision was made by an individual or group of individuals that the company was not going to allow to fail.

With a Herculean effort on the part of the Mississippi Operations, and with a lot of help from up North, harnesses went out the door in sufficient quantities and timing to keep from shutting down vehicle assembly plants, which would have been a truly unforgivable offense. But, it didn't take long to convert the operations back to tried and true technology. It was just sad to again see such a blunder by supposedly smart people. Toyota would have never done it. Are you seeing a pattern here?

In addition to the obvious errors in judgment that were made, another critical fallacy was made that was often repeated in the low labor cost Mexican operations. That fallacy is the belief that if you are operating in a low labor cost operation, it is not all that important to tightly control labor. Merely focus on the other things like investment, inventory, transportation, etc., and let labor take care of itself. According to some: "It just won't amount to enough money to worry about." Nothing could be farther from the truth.

The naïvety of this attitude is just astounding. Even in Mexico, with $2/hour, fully burdened labor costs (and this was some years back, it is much higher now), this is a very foolish attitude and shows a total lack of understanding of the cost profiles that exist with manufacturing plants.

First of all, if you do not control labor, it is highly unlikely that anything else will be controlled in a manufacturing environment, such as quality and delivery. There is a direct relationship between quality and delivery and productivity, and anyone who has actually worked in first-line supervision knows this exceedingly well. Another fact is that even in an environment with very low labor costs (relatively speaking), the cost for direct labor is still by far the biggest expense for any plant, excluding the cost of materials, which is marked up when included in the price of the product and, therefore, not an expense in the same vein as other plant expenses.

On a Packard Operating Report, most of the expenses incurred by a plant in the performance of its duties were listed, along with the cost of each of items, such as direct labor cost, indirect labor cost, salary labor cost,

cost of supplies, depreciation, spare parts, cost of floor space, utility costs, repair and scrap costs, transportation, travel costs, some allocations, etc. On a typical Operating Report (department budget), about 70% of the total expenses were due to labor and labor-related costs, and, of this 70%, usually all of it was for direct hourly labor.

In other words, about half of a plant's operating budget was for direct hourly labor, and this was in Mexico where the cost of labor was (used to be) very low. The other fact, not to be dismissed, is that virtually all of a plant's other expenses are heavily impacted by the amount of direct labor in the plant. Each operator needs supplies and a workstation, which requires an investment, uses utilities, and takes up space. Indirect operators and salary employees are in large part determined by the number of hourly direct employees. Somehow, much of this had been lost on Packard upper management.

Most Packard executives were more than anxious to implement so-called Lean Initiatives to save a few labor dollars, especially in more recent years (initiatives that were often anything but truly Lean, as we will talk about in the next chapter). But, getting back to real productivity improvements through tried and true processes and controls never seemed to be a priority within Packard.

Integrated Production System (IPS) to the Rescue (or Not)

The taste was bitter enough from the Chevy B FL fiasco (see above) that no further automation attempts were made for the next few years. However, it was well known throughout the organization that the general manager was hot-to-trot to see some major changes in our wiring harness manufacturing technology, and he kept using the word "automation." I believe it was in 1977, just prior to when I was transferred to El Paso as part of the Mexican Operations startup team, that a new group was created that was tasked with "exceeding the expectations" of our managing director regarding the creation of new wiring harness manufacturing technology. I was ecstatic that I was going to Mexico and not being asked to be part of that team.

The truth is that it was a misguided effort from the start, but no one wanted to tell that to the general manager. There was just no way, without *major* harness redesign (a totally new concept of providing electricity throughout the car), that automation was going to work. And, if automation was totally unjustifiable with the very high labor costs in Warren, Ohio, it certainly wouldn't be justifiable anywhere else. So, if one is tasked with running this new team, what do you do, especially if you are an empty suit running the show? This was a very unenviable position, to be sure.

The group was astute enough to know that automation, as had been tried before, would result in utter failure. They also knew that they had to come up with something that looked different than what was currently being done. They also knew that a catchy acronym for whatever they came up with would be a real plus.

They also were smart enough to realize that they needed to choose the right song (the right product package), one that gave them a big improvement potential. That is, choose a product in current production that had not been preplanned as well as could have been (or had been in production several years and had not been repreplanned to accommodate all of the engineering changes), and one, therefore, that provided an opportunity to obtain savings by just doing a better preplan job, regardless of what the rest of the system entailed. They found just such a product in the plant: the high-volume Chevy B Instrument Panel—and they went to work. By the way, the IP harnesses are the largest harnesses in the car with four or five times more work content than any of the other harnesses. This would mean that there was more opportunity to realize labor savings and more content on which to spread any new investment requirements.

After a lot of careful considerations, they came up with a catchy name for their new manufacturing system: Integrated Production System (IPS).

There were a couple of new aspects to this system that were rather interesting. One concept was a redesign of the harness build fixtures (they called them "low profile" boards) such that the height of the build fixtures (boards) was significantly reduced making it easier for the operators to reach the higher plugs. In fact, the fixtures were low enough that it was possible to put some of the connectors in bins above the build fixtures in the center of the line so that an operator would not have to turn around to get a part (see sidebar).

I should explain a little about the conveyor system so that I do not lose you. Packard had several different types of conveyors, but each of them preformed basically the same function of transporting build fixtures (boards) on an oval track or platform past the production operators at a predetermined speed, in order to meet customer delivery requirements. The boards were mounted, with a consistent spacing between boards, on parallel tracks or platforms such that there was only a foot or two between the backs of the boards traveling down one side of the line and those traveling down the opposite side. Boards reaching the end of the line would make the turn and then travel back down the opposite side from which they came.

In the Methods Lab preplanning process, the required line speed would be determined based on the output requirements, which would then dictate the number of operators required based on the total work content as determined in the preplan process. The work content would then be divided evenly among all of the operators in the most efficient sequence possible. The first operator on the line would build his/her content onto an empty board, the second operator would add to what had been done by the first, etc. The last operator would complete the harness build, perform an attributes inspection as he/she would take the harness from the board (the harness would have already been electrically tested on the line), and then fold the harness and pack it into a container. An empty board would then arrive in station #1 and the process would begin again.

The second new concept pertained to how cut wires were handled and inventoried from the cutting department. Generally, cut leads were put into plastic totes by the cutting operator and moved to the final assembly area, then the leads were put into predetermined storage locations in the Final Assembly Department storage racks by a service operator. When a wire type was needed on the conveyor, a final assembly service person would go to the storage rack, get a bundle of the needed wire, carry the bundle of wires to the line, and service the wire to the appropriate rack in a manner such that the conveyor operator would avoid tangling wires when getting one wire at a time. The new concept entailed having the cutting operator (who had some dead time within the cutting operation) put some of the lead codes, which had the right characteristics, directly onto specially designed racks instead of putting them into plastic totes. Then a service person would wheel these racks of leads directly to the line.

To some, this sounded synchronous and that it ought to work great, but I and others knew the logistics were going to be a nightmare, take up a lot of valuable space, and most likely cost much more than could possibly be saved.

The problem is that what they had come up with so far was not different enough from current technology to impress the general manager, so, they came up with a new concept that did look a lot different. On a typical conveyor, a set of racks, trays, and bins would be positioned behind an operator that would hold all of the materials he/she would need to complete his/her build station. A service person would be assigned to the line who would ensure that the operators did not run out of materials. As soon as a rack or bin would get low on material, the service person would get additional material from the storage area and replenish the line (this is a pull system even though many Packard managers didn't recognize it as such).

What the team came up with was the novel (not good, just novel) idea of putting a service conveyor around the harness build conveyor (in front of the operators) at a height equal with the bottom of the conveyor platform. This service conveyor would be twice the length of the build conveyor. The materials that would be placed on the service conveyor would be placed inside or near the half of the service conveyor that would extend past the harness conveyor.

The idea was to have the service operator load this half of the service conveyor with materials while the build line was using materials off of the other half of the service conveyor. Then, at a break or at lunch time, the service conveyor would be indexed so that the fully loaded half of the service conveyor would be appropriately positioned in front of the operators, and the mostly empty half of the service conveyor would be in position to fully load again, in time for the next rotation.

I hope I have explained this well enough so that you can understand what this concept was all about. But, like Winston Churchill said, "There's nothing wrong with change, as long as it is in the right direction." Well, this wasn't a change in the right direction. The investment in tooling and equipment was going to be over $200,000, not to mention the additional floor space requirements and what additional maintenance was going to be required for the new service conveyor (my guess was that it was going to be substantial). But, the biggest question for me was: Where are the savings?

There would be a little savings in the reduction of walking time for the service person (depending on the congestion), but I wasn't sure that would offset the built-in inefficiencies of this new service load. Depending on the size of the various racks and the frequency of the service conveyor rotation, many circuits and components would not be serviced based on the most convenient and efficient cycle. Many of the wires were not well adapted for the low height of the service conveyor and would get tangled and fall on the floor as the line was moving or the line operators were getting a lead, which would require the service person to fix them or the operators to pick them up. In addition, many of the wires were not well suited for the operators to get from low racks (resulting in labor penalties and not the slight labor advantage you would normally get by eliminating a turn).

Well, none of this stopped the team from proceeding post haste. Fortunately for them (but not for Packard), the harness they had selected was poorly balanced from the prior preplan job, and they were able to repreplan it and show a forecasted savings of about $250,000 for the entire IPS. Because they were only going to spend about $200,000 for new tooling and equipment, they were touting a payoff of only 0.8 years for this new system ($200,000 cost/$250,000 yearly saving = 0.8 years payback).

Generally, any investment with a payoff of less than one year is considered very good and is almost always approved. Because the general manager liked this new concept, there was really never any doubt about it being green-lighted.

I was in El Paso at the time that most of the IPS project was in the planning stages in Warren, busily involved in the startup of our first Mexican plant in Juarez, but I was making frequent trips back to Warren to participate in forward planning meetings and in observing the progress of the preplanning work being done on the harness packages that were targeted for our plant. During the trips back to Warren, I had occasion to view some of the work being done in the lab on the IPS program, and after seeing what they were planning, I had some real reservations. Once I understood what they were doing, I just couldn't see any possible way that this system could be justified. It was obvious that the savings were totally attributable to fixing a poor prior preplan job, and these savings could be obtained without spending any of the $200,000 costs attributable to IPS (which costs, I had strong inklings, were grossly understated).

One of the things I had been taught in the Industrial Engineering (IE) Training Program is that anytime improvements are being contemplated to any process, which will require an investment of any significant amount of money, the first thing that must be done is to ensure that the "best manual method" (or best "low-cost manufacturing system") has been established as the baseline. This must be done so that a realistic determination of the potential savings of the proposed new system can be made. This is only common sense. It was quite clear from what I had observed that this was not what was going on.

To confirm my suspicions, I arranged a meeting with the supervisor of this project, who reported to the "suit" I referred to earlier. I asked him to explain the entire concept to me in detail, including the cost and the savings breakdown of the system. He was very reluctant to divulge the breakdown of the savings, but, with enough arm twisting, he finally conceded that the savings attributable to the service conveyor and the new concept of servicing wires onto special racks in the cutting area accounted for only about $10,000 worth of savings (and I believe he was exceedingly optimistic). The rest of the savings, about $240,000 per year, was attributable to the redesigned low-profile boards and the repreplan of the line.

I asked him, in light of this, how he could propose proceeding with the service conveyor, especially since very optimistic projections had been made regarding its dependability and its long-term maintenance requirements.

The concept of having the cutting operators service wires directly onto special racks and having these racks pushed into the Final Assembly Department also seemed very questionable, in light of the obvious logistics problems and the minimal savings (negative savings?) that resulted from it. Regarding the low-profile boards, because it didn't cost any more to build them than it did to build the existing type of board (when there was a need to build new boards for a harness package), low-profile seemed like the right way to go, and, obviously, the new preplan job, which was responsible for almost all of the savings with none of the cost, should be implemented.

His response was that this was a system and needed to be installed in entirety. The fact is, he didn't believe this any more than I did, but he was in a terrible position. There were very few people I knew at Packard who would have made the right, but probably career-ending, decision to come clean about this system if they had worked on it and knew the general manager wanted it. They had worked for months on this project and had utilized a lot of resources, human and otherwise, to come up with a new system. And, most importantly, the "suit" wasn't going to back off of the project because he knew the general manager liked it, or least, he liked what he had heard about the totally new concept and the savings involved.

At the end of my meeting with the IPS supervisor, I said something like: "So, you are going to support, or at least not challenge, a decision to conservatively spend $200,000 to save a somewhat optimistic $10,000, instead of supporting a decision to spend nothing (the preplan job had already been done) to gain a savings of $240,000." His answer was that he really didn't have a choice.

I disagree. He did have a choice, but if he did the right thing it could (or would) have been a very damaging one for his career. Upper management at Packard did not like to have lower level employees disagree with them, and when it happened, it normally didn't work out that well for those brave enough to do so. If you read John DeLorean's book, *On a Clear Day You Can See GM* (Wright Enterprises: Books, 1979), it seems that this is a characteristic of Packard that was inherited from its parent.

I had discussions with my boss back in Juarez, but nothing happened. I didn't really expect it to. He wasn't interested in sabotaging his career, either. The line was ultimately installed in the Mississippi Operations, much to their chagrin, even further lessening any positive impact this system might have had, since Mississippi was still a low labor cost location at that time. Also, Mississippi had much less Process Engineering expertise than did Warren, so it was really questionable to me why they would choose to put the trial run into Mississippi instead of Warren. There are only two reasons I can imagine: (1) there may have been a concern that the union would have a negative influence on the success of the system, but because Packard was trying to reduce labor in the wiring harness manufacturing process in order to keep Warren viable, this doesn't seem likely; and (2) they might have been trying to get the system as far away as possible from Warren headquarters so the big bosses wouldn't have a daily reminder of the mess they had foisted on the organization.

Mississippi really struggled with this line. By now, there had to be a lot of people besides me who were questioning this decision. The service line was constantly down and required a lot of maintenance that was not forecasted. The logistics of the new system to service wires directly from the cutting area was creating nightmares. Wires serviced to the service conveyor were getting tangled, some falling to the floor, and, in general, not creating the supposed advantage to the operators on the conveyor line. And, it was very ineffective to service wires to the service conveyor, because partially full racks were coming back at the end of each rotation, and often the racks contained tangled wires that required additional labor to straighten. All of these things were easily predictable.

I'm sure what most of you are thinking is that it became obvious to everyone that this system was a disaster that should be put on the Packard trash heap of history, never again to be resurrected. Well, you would be wrong. Unbelievably, and I use that word in the strongest possible terms, a decision was made to expand this system and to install it on a total of 16 conveyor lines.

How could this happen? I can't imagine any sane person making this decision, but it was made. Again, I am sure it can be traced back to the fact that our general manager thought it looked like a cool concept and he was told it saved money. Did the Mississippi leadership speak up and say the system isn't working? I don't know who said what to whom, but the end result is that one of the worst concepts in the history of wiring harness manufacturing not only failed to be terminated, it was dramatically expanded.

This expansion was taking place in the late 1970s and early 1980s and things only got worse as time wore on. The service conveyors began to wear out after a year or two and it got to the point where there were maintenance people crawling all over the place trying to keep them running. Large sums of money—money that Packard was trying desperately to save—were going right down a rat hole, and Packard really could not afford the luxury of a blunder of this magnitude at this time.

It finally got to the point that something had to be done or most of the Mississippi operations staff were going to end up greatly frustrated. It took one of my good friends, who had a strong methods and IE background, to come up with the perfect strategy. He was able to sell the idea that Mississippi, in the spirit of continuous improvement, had come up with a great enhancement to IPS, and, to make matters even better, they had come up with a great acronym: HIPS, or Hybrid Integrated Production System. Mississippi realized that they had to kill IPS, but to do it in such a way that the General Manager and the other leading high-level supporters did not "lose face" (a concept I learned a lot about in South Korea).

What was HIPS? It was IPS without the service conveyor and the special racks and the special handling of wires from the cutting area. In other words, it was the manufacturing system we had been using successfully for decades, except with boards of low-profile design such that some materials could be placed in the center of the line above the boards (and even the low-profile design, with materials above the boards, had been used successfully in the past on several smaller harness types).

With this new strategy, Mississippi was finally able to put a stake in the heart of this disaster, never again to be resurrected. One had to believe (or at least hope) that Packard had learned its lesson, and that no one else would ever be foolish enough to try it again. Although, for competition sake, Packard could hope that some competitor would be naïve enough to "benchmark" this total failure from the world's biggest wiring harness producer.

Therefore, what was the final tally after having this system in place for only a few years, which must have seemed like an eternity to those forced to try to make it work? The conservative estimate that I heard was that it cost about $1 million per line over and above the cost to produce the same harnesses on a conventional system ($16 million in total). I suspect that in reality it cost a lot more, as it is very difficult to capture all of the costs associated with a system like this, not to mention the damaging effect on employee morale. But, even if the million dollars per line is accurate, everyone had to understand that if Packard kept making mistakes like this, it wasn't going to be too long before it started adding up to real money (money that Packard and its parent, GM, just didn't have to waste).

By the way, the empty suit had been moved to another exciting assignment. Unfortunately, it was an assignment to the Mexican Operations where he would soon replicate his history of "sterling successes" long before it "hit the fan" on IPS. When the dominos started to fall, it was hard to find anyone who even remembered he was largely responsible for this debacle, along with the general manager and his yes men.

When I wrote and compiled an IE Manual and associated training program some time later, I added an addendum to the concept that all new systems being proposed must be compared to the "best manual method," or to the "best low investment option" in order to determine true potential savings. This addendum stated: *When a new, high investment project is being proposed, every element within the system, which can be feasibly separated from the total system with no negative impact, must be justified on its own merits.*

If this concept had been incorporated by the IPS team, Packard would not have made the horrible decision to spend $200,000 (very conservatively) to save $10,000 (very optimistically).

If IPS Didn't Get It Done, IPS II Surely Would

IPS was just in its infancy (around 1979) when the decision was made to further explore the type of automation (semiautomation) that had already been tried, albeit, with utter failure as described before on the Chevy Impala Forward Lamp package. Our general manager was bound and determined to make automation work. This time Packard was not going to make the same mistake they had made before by selecting the wrong harness type, but, that was only part of the problem, as you will remember. Instead of selecting a long and difficult-to-handle forward lamp harness with two big splices tying most of the harness together, the decision was made to select a relatively simple engine harness. Engine harnesses have similar work content to the forward lamp harnesses and, because of common platforms within GM, the volumes for engine harnesses were relatively high. Engine harnesses also were much shorter and easier to handle because none of the wires (leads) were tied together by complex splices.

Technology had not changed in the six or eight years since the earlier Chevy FL debacle as to the type of equipment and tooling that would be placed on the line to cut and terminate the wires (leads). But, a new type of conveyor was incorporated with this system called a *power-and-free* conveyor, which worked in conjunction with a conventional conveyor, much like the service conveyor meshed with the conventional final assembly conveyor in IPS. The first part of the power-and-free conveyor would contain pallets with holding fixtures, which would hold the leads as they were cut and terminated automatically by the equipment. Then the pallets would pass in front of the operators on the second part of the conveyor, who would each take the leads destined for their individual station and build them to a low-profile board on the conventional final assembly conveyor.

To some of you reading this, this may sound like a pretty neat concept. But, much like the question I asked about IPS: "Where are the savings?" If all went according to plan, there would be some savings in not having to service the automatically produced leads to the line, as well as eliminating the labor to cut and terminate the leads on conventional cutters. But, what of the extra maintenance personnel and engineering resources that would be required, and what about the hundreds of thousands of dollars of extra tooling and equipment and floor space that would be required for each line? Some felt that the final assembly operators would realize some savings by having some of their leads located in front of them as the power-and-free conveyor passed by, but that savings would be very minimal, if it existed at all, due to the nature of how the leads would be presented to the operators and the potential problems associated with synchronizing the power-and-free line with the build line.

And, the savings were projected based on everything going according to plan, which was that the system was going to run at an uptime percentage in the high 90s. Some might call this a pipe dream. Based on prior experience, I would call it lunacy.

However, I would not quarrel too much if Packard upper management made a decision to make an investment of about $1 million on an evaluation line in an effort to gain more knowledge and possibly develop some new technology, and, perhaps, save a few dollars in the process. The problem is that Packard went into production in an actual manufacturing plant long before it was ready for prime time, and the decision was made to proliferate this technology on all engine lines before any meaningful evaluations had been done. Based on my experience at Packard, once the die was cast, there was no turning back, regardless of what future events might unfold. Again, this is something that Toyota would have never done in a million years.

I followed the progress of IPS II from afar, because its success or failure could have an impact on the Mexican Operations; the reason was that Packard had already made a decision that the next harness type to get the IPS II treatment would be the Forward Lamp. I was working on forward planning for Mexico and we were busy planning the startup of the second plant.

We had started up the first plant with Rear Body Harnesses, and the startup had gone very smoothly. Forward Lamp Harnesses were similar in design and used similar components, and we were hoping we could obtain this package for the second plant. The IPS II pipe dream could definitely throw a monkey wrench into our desired plans.

I talked to the Engineering director in Warren and told him of our plans, and he said that we would have to change them due to IPS II's long-range planning. I knew that IPS II would be the biggest bust of all, so I was not worried that Forward Lamp Harnesses would ever see IPS II. Even with recent history, there was just no way Packard management would be so stupid as to proceed with it after the Engine Program would undoubtedly crash and burn. I told the director that we needed the Forward Lamp Harnesses, which were similar to Rear Body Harnesses, to ensure a smooth startup in the second plant. However, I assured him that when Packard was ready to proceed with Forward Lamp Harnesses on IPS II, we would be more than agreeable to transfer them back to Warren in exchange for another program.

Believe it or not, the ploy actually worked. For a long time, I thought it was because of my great salesmanship, but, upon reflection, I don't think he had much confidence in the IPS II forward plan either, especially since the first engine line was running an uptime percentage in the 50s. Because of this, backup leads had to be cut and terminated on standard cutting machines and serviced to the line in the conventional manner on racks behind the operators. With tremendous effort and dedication, the uptime percentage on the IPS II line got into the 60s, and maybe touched the 70s occasionally, but, in total, it was a totally unreliable process requiring redundant operations throughout the department in order to keep Engine Harnesses going out the door.

So, this was the end of it, right? Well, no. I believe that five or six lines were ultimately installed (it might have even been more), each performing about like the first. Again, Toyota would have never installed the first line into a manufacturing department, much less a half dozen or more. But, miracle of miracles, the decision was finally made to cancel the deployment of IPS II, after the Engine Lines clearly demonstrated again that this idea just was not going to work.

But, again, how could a decision be made to proliferate this technology, which was not proved and which had very questionable justification? Toyota would never have done this. The additional maintenance men who were constantly in the area more than offset the few production operators who would be eliminated if the system worked as planned, and how did we ever hope to pay off an investment of hundreds of thousands of dollars in tooling, equipment, and floor space for each line? It was truly a pipe dream.

Also, how were the estimates made that were used to "justify" this system. I talked to one of the engineers who had been working on this project from the get-go. Our conversation went something like this:

My Question: What kind of an uptime percentage did you estimate for the system?
His Answer: We felt that it would run in the high 90s.
My Question: How many pieces of equipment are there on the line and what is the expected uptime of each?
His Answer: About 10 (*to be honest, I can't remember the exact number, but there were a lot*), and we expect each to run an uptime of between 95 and 98%.
My Question: Are you aware that the uptime of the system is approximately equal to the uptime percentages of all of the pieces of equipment multiplied together?
His Answer: Really?

I'm afraid that this is just about how it went. Did no one on the team know how to calculate a process uptime? Surely they did. I suspect that the empty suit didn't want to know the answer; the general manager wouldn't like it. By the way, if there were 10 processes on the mech line, each with an uptime of 98%, the uptime of the system would be a little over 80%. Because Packard never reached these lofty heights, some of the individual pieces of equipment must have had uptime percentages in the mid-to-low 90s.

Packard might have thrown away several more millions of dollars, but at least we got to follow our desired forward plan in Mexico and start up the Forward Lamp Program in the second plant, with no give backs.

Remote Lead Prep: Yeah, That's the Answer!

By the end of 1982, Packard had successfully started up three full-service plants in Juarez, in addition to two full-service supplier plants. A full-service plant is defined as a plant in which all of the wire cutting and terminating, as well as all other lead preparations, are done within the plant in support of the final assembly operations. When we were planning the startup of the first plant, it never entered our minds to consider anything other than a full-service plant. Of course, we received all bulk cable, terminals, connectors, and other materials from Warren, but, the cutting and lead preparation operations must be done within the plant; of that, there could be no doubt.

Without question, wire cutting and other lead preparation operations require more skilled workers than typical final assembly operations. Cutting operators can generally be trained and become efficient in a matter of four to six weeks, and, depending on the lead prep process involved, lead prep operators are generally efficient in four to eight weeks, whereas, a final assembly operator is normally efficient in a couple of weeks. But, with a little bit of selective placement and proper training, it is not difficult to get a Cutting or a Lead Prep Department up and running. And, as long as the startup curve is understood, plans can be put in place to ensure that sufficient material will be available for the final assembly startup, with no risk of having insufficient materials to run continuously.

Maintenance requirements also are somewhat more complex in cutting and lead prep operations, in order to keep the cutters and lead prep equipment running and the application dies adjusted and repaired as needed. This was all taken into account in preparation for our first plant startup. There were four engineers who were part of the startup team for the first plant. I had responsibility for Industrial and Methods Engineering, a second had responsibility for all tool and die issues, a third had responsibility for all cutting machines and other lead prep equipment, and the fourth had responsibility for all electrical issues in the plant.

I also had the responsibility for training lead prep and final assembly operators, and, with the help of a couple of operators from Mississippi with lead prep experience (who also conveniently spoke Spanish), we were able to provide effective training to the Mexican lead prep operators in methods and quality considerations. All three of my compadres did a great job of finding and training Mexican engineers and technicians, who in turn provided additional training to the operators regarding station setups, die changes and adjustments, and other important aspects of their jobs. The end result was that the startup of the cutting and lead prep areas went very smoothly.

The second and third plants started up much like the first, even easier in some regards as we had a good base of experienced Mexican engineers and technicians who could provide support to the new plants on an as-needed basis. That's not to say that the cutting and lead prep operations were without problems. There are no manufacturing operations without problems. However, in total, the concept of full-service plants was very successful and there were very few, if any, episodes where final assembly lines had to be shut down for lack of material, which is the cardinal sin in final assembly manufacturing.

Sometime in mid-1983, a couple of Warren engineering managers were sent to Mexico to present a new concept of cutting and lead prep to the Mexican operations. The new idea was to scrap the very successful concept of full-service plants in favor of receiving all future cutting and lead prep requirements from Warren, Ohio.

We told them in no uncertain terms that this was a horrible idea. Many others were as opposed to this as I was. We knew without a doubt that if this plan was implemented, it would be a disaster for the Mexican operations, and it would dramatically increase overall costs for Packard. It would add millions of dollars of unnecessary inventory into the system and it would create a 2000-mile supply line for very sensitive materials. We knew if this idea was adopted, our final assembly lines would be up and down like yo-yos, which would have a very negative impact on the productivity within our plants and the attitudes of our employees, which would be manifest in other areas as well.

What came out of the meeting from our side was pretty much: "You'll do this over our dead bodies." And, this was a universal sentiment. After all, Packard was becoming very infatuated with Synchronous Manufacturing, in theory at least, and there was nothing less synchronous than having cut leads coming from over 2000 miles away.

The problem was that we just didn't know that the decision had already been made and Warren was just being polite in asking our opinion. For anyone reading this book, please do not do this. If a decision has been made, just inform the appropriate parties of such. Don't pretend that what they say can affect the outcome when it can't. It just depresses people and makes them question your leadership skills.

A few weeks after we had sent the Warren managers back home with not only a "no," but with a "HELL NO," they returned with the information that we had come up with the wrong answer—remote lead prep was to be the new winning strategy for Packard.

I just couldn't believe it. Sure, Packard had done a lot of stupid things in the past, but this? Of course, we asked why and we were given two reasons: (1) Packard intended to convince the union that in order to keep

Packard viable, Warren would become the high-tech manufacturing location for cable, components, cutting, lead prep, etc., and Mexico would be the low-tech manufacturing location for final assembly. The thought being that this would make it more palatable to the union for Packard to keep sending more and more jobs to Mexico, and (2) because it was somewhat easier to start up final assembly processes than either cutting or lead prep; jobs could be moved faster to Mexico if Warren did all the cutting and lead prep and Mexico only had to worry about final assembly.

Neither of these reasons held water as far as most of us were concerned. Packard wasn't terminating any union employees as a result of the Mexican Operations. New business was being moved to Mexico and current business was being moved out of Warren only as attrition allowed.

It was clear that a certain number of jobs would remain in Warren, regardless of what happened in Mexico. Surely, it made sense to keep cable making and component making in Warren, but as to cutting and lead prep being high tech, this was nonsense. What did make sense was to utilize the balance of Warren labor (over and above cable making, component making, and a few other relatively high-tech jobs) in the most effective manner possible.

For example, the Lordstown Assembly Plant was located only about 20 miles from Warren, but the harnesses for this plant were mostly built in Mexico, over 2000 miles away. Several other assembly plants were located within a few hours drive of Warren, but these plants were two or three truck days away from the Mexican plants. In other words, why not utilize Warren to supply wiring harnesses to Lordstown and use the remaining Warren labor force to build wiring assemblies for other plants near Warren, and use the Mexico plants for all other wiring harness requirements?

I wasn't sitting in the meeting with the union talking about Packard's strategy of eventually shipping most of Packard's wiring harness business to Mexico, so it is easy for me to say what should have been done. Maybe it might have been easier to convince the union of the new strategy that Packard wanted to pursue (remote cutting and lead prep) than it would have been to convince the union of the need to properly rationalize the business, as suggested above. But, I cannot believe that it could not have been done. What Packard management should have known is that a strategy of remote cutting and lead prep was going to be very expensive, and a gigantic nightmare.

Shortly after the decision to transition to remote cutting and lead prep, I was asked if I was interested in taking an assignment in Packard's European Operations, as the engineering manager for the Portuguese plants.

Under normal circumstances, I don't think I would have been all that excited about spending several years overseas. However, when I thought about all of the pain and misery the new lead prep strategy was going to inflict, the offer started to sound pretty good.

I made the decision to take the job, and it turned out to be a very good decision. Not just because I really enjoyed the international experience, but because I avoided the mess that was created by remote cutting and lead prep. Everything that we projected would happen, in fact, did come to pass—and it didn't take long.

The productivity of Packard's Mexican operations at the time I retired in 2001 continued to hover around 50% of the level attained in Warren in the early 1970s, when all of the tried and true basics were still in place. There were four primary factors that were responsible for this: (1) a Mexican turnover rate of up to 100% or more per year, (2) the fact that manufacturing departments were never organized properly as had been established by decades of success in Warren (see Chapter 5), (3) the disintegration of the system of measuring and controlling productivity (see Chapter 6), and (4) remote cutting and lead prep as described in this section.

How about Just Getting Rid of Our Productivity Control System

I'm only going to touch on this disaster (getting rid of our productivity control system), because I will be devoting all of Chapter 6 to this subject. However, it certainly bears mentioning in this chapter listing some of Packard's all time horrible decisions (not counting the ones related to Lean covered in the next chapter).

Before I started at Packard, I had the opportunity to work summers in other manufacturing companies. I also have had the opportunity to observe many other manufacturing companies over the years, and none has had a really good system to control efficiency, which is a very large component of productivity. But, when I started with Packard, I found out that they did have just such a system. It had a few weaknesses, which wouldn't be that hard to fix, and it was based on very old technology (1971), but the concepts of this system were very sound; much better than anything else I had ever seen or have seen since in other companies. This system was largely responsible for the very high efficiency levels being run in the Warren Operations.

Anyone who had worked in both engineering and manufacturing knew how indispensable this system was in helping a foreman run an effective manufacturing department. Sure, the system needed to be given an overhaul, but the basic concepts were rock solid and very critical to effective operations.

In 1984, an academic with a background in physics, Eliyahu Goldratt, wrote the book *The Goal* (North River Press, 1984). This book created quite a stir within Packard. I was in Portugal at the time, but I was sent a copy and asked to read it. The book was written by Goldratt over a two-year period and was written as a novel. It detailed the efforts of a manager of a very poorly performing plant to try to save his job, as well as the jobs of everyone else within the plant, because the plant was on the chopping block. Unless there was a quick and dramatic turnaround, the plant would be gone.

The book was an interesting read, but I had a hard time understanding what the fuss was all about. Packard had been around a long time and had learned a lot in the process. There were some good basic concepts presented in the book, but all of these concepts were already known and practiced within Packard. Packard already understood the importance of throughput as opposed to just creating inventory; the need to put bottlenecks at the end of a manufacturing process and have excess capacity upstream; the importance of using pull systems; the need to balance flow with demand, as opposed to capacity; along with several other basics.

There were a few very questionable statements/recommendations that were made in the book, but I didn't think too much about them. They wouldn't have any bearing on how Packard did business, or so I thought. There were contentions in the book that a plant with everyone working all of the time was a very inefficient plant; that excess capacity should sit idle (meaning the people and not just the equipment); that batch sizes should be drastically cut on nonbottleneck operations; that time saved on nonbottleneck stations is a mirage.

These recommendations/statements only made sense if one assumes that a person is assigned to every work station within the plant and that these workers are not moved to other jobs once their initial assignment is complete. Does anyone really operate a manufacturing operation like this? I know that Packard surely didn't.

By the nature of our business, it frequently (in fact, always) occurred that a foreman would have one piece of equipment, but would only need 0.4 operators to meet demand, or would have 2 pieces of equipment with a need for 1.3 operators, or would have 5 pieces of equipment with a need for 4.1 operators. A good foreman would move the operators around as needed to meet the demand and ensure that every operator had a productive job to do at all times. If a Foreman temporarily had more operators than necessary, he would loan them out to another foreman who was short of help or he would find something productive to do with them, even if it meant just cleaning and organizing the department. If he was temporarily short of help, he would borrow from someone who had a temporary excess or he would schedule overtime. *The one thing he would never do is allow an operator to sit idle.* It doesn't take a rocket scientist to understand why this is not a good idea.

But, like I said, I wasn't too concerned about these recommendations, because Packard had long since learned how to deal with excess capacity and run efficiently at the same time. I couldn't have been more wrong.

When I came back from one of my home leaves from Portugal, I found out that Packard had terminated its long-held and very effective system for controlling productivity. Packard didn't make the logical and correct decision to enhance the system through utilization of improved computer technology so that it was even more useful, it just flat out cancelled it. I couldn't believe it. Who had made this decision and how had he been convinced that it was the right thing to do?

Apparently, Goldratt's book had convinced a lot of Packard upper management that the only things that mattered anymore were throughput and profitability. Efficiency just wasn't something to be concerned about, because focusing on efficiency would probably lead us to make bad decisions in other areas, like quality. Little did they understand, apparently, that without efficient operations, you will get neither acceptable throughput nor profitability over the long haul, and there is no conflict between efficiency and quality in a well-run department. In fact, efficient, well-run departments put out the best quality (just ask Toyota).

From that time on, when ever anyone was touring a plant and an operator was observed sitting idle, the response from manufacturing supervision was: "They do not have a schedule. You don't want me to build unneeded inventory do you?" It was as if this was the only possible option. How about managing your people effectively so that you do not create excess inventory or have people sitting idle? Both are certainly possible. It had been done effectively at Packard for decades.

Getting rid of the efficiency control system was one of the worst decisions Packard ever made because of its long-term consequences, right along with remote cutting and lead prep. Most of the other disasters I have discussed in this chapter only went on for a few years, and then the pain was over. The pain from this decision continues to this day.

So Just How Are IDCs Going to Help Us with the 3% Give Back?

I returned from South Korea in 1990 and was assigned to the Mexico East Operations headquartered in Laredo, Texas. This operation had been initiated and built up during the seven-year period I had been overseas. Shortly after I returned, I learned of an edict that GM had given to Packard, and each of the other vertically integrated GM family companies, to the effect that each would be compelled to give GM price reductions for each of the products sold to GM each year for the next several years. It seems the first-year price reductions were scheduled for 5%, or maybe it was 3%, with an additional percentage being tacked on each succeeding year. This probably wasn't as big a problem for Packard as it was for the other GM subsidiary companies due to the positive impact of engineering changes on our pricing and profitability, but it still presented a challenge.

The first assignment I was given upon returning to the States was to provide engineering support to two new plants being started up in Victoria, Mexico, which was located about four hours south of Monterey. In one of the plants, a new component was being introduced called an Insulation Displacement Component or IDC. Its purpose was to replace conventional splices in the wiring harness, which had gotten a bad rap by some in upper management. In reality, splicing was relatively trouble free and very inexpensive.

The IDCs were made with two halves, each half being comprised of a plastic housing that contained a slotted metal strip. The IDCs functioned by having the operator place each of the wires comprising the splice into one of the slots of the metal strip on the bottom half of the IDC, placing the top half of the IDC over the bottom half, and then closing the IDC with a small press. The act of pressing the two halves together forced the wires into the metal slots, causing the insulation on the wires to be cut so that the copper wires made contact with the metal strip, thereby completing the circuit for all of the wires within the IDC.

My understanding was that studies had been conducted and it was determined that an IDC would save money versus a conventional splice, although I was never able to find the studies or find out who had done the studies and under what conditions. Based on my extensive IE experience, I was extremely skeptical that IDC would save money or improve quality, but the decision had been made and we were going to have to make it work.

What we found out once we got into the preplanning program is that my suspicions were well founded. We not only failed to save labor with the IDCs, we, in fact, had a very significant cost added to apply an IDC on-line, versus making a conventional splice either on- or off-line and then building it to the board.

This was not all. We had a real concern about quality because it was difficult to get consistently good continuity. This was never a problem with standard splices. The biggest quality concern I had was that it would be very possible (more like probable) to get continuity on the board so that the harness would ring out as good, but then lose that continuity in the course of shipping that harness to the assembly plant and/or building the harness into the car or, even worse, as the car was driven down the road. We never had this kind of problem with conventional splices, but we did experience this problem with IDCs.

Lastly, the IDCs were fairly bulky, such that when all of the wires were built into a harness, including those that went into the IDCs, and the harness was covered with conduit or with tape, it looked like there were giant tumors in the harness (kind of like when a snake swallows a mouse). This certainly didn't give the impression of good quality.

The bottom line is that this was another bad decision by Packard based on a poor (or, more likely, a nonexistent) study, and, more importantly, a decision to go into production with an unproven process. Toyota would not be happy, again.

The funny thing is that when I laid out all the facts for my boss, he indicated it would be best to be mum about it. Apparently, there were some big toes that didn't need to be stepped on. My problem is that I'm not much of a diplomat. When the Warren Manufacturing director came to our plant to see the progress of our startup, he asked me what I thought of the IDCs, so, I told him. He seemed shocked and asked how Packard was going to offset the 3% give back to GM if IDCs didn't save any money. I told him I had a lot of ideas I would like to share with him, but none of them included the introduction of new technologies that increased costs. By looking at his face, I got the impression that somehow he had been involved in the decision to proceed with this idea.

How about Let's Just Get into a New Business

Peter Drucker spent a lot of time at GM in the 1940s, but you could never tell it by the impression his ideas had (or, more appropriately, didn't have) on GM and the GM subsidiary companies. Drucker contended that companies tend to produce too many products, hire employees they don't need when outsourcing would be a better idea, and expand into economic sectors they should avoid. In the mid-1990s, this would have been very good advice for Packard to have heeded.

Hughes Electronics, which was part of the Hughes Corporation, which GM had purchased, was assigned to Packard Electric in the mid-1990s, because their products seemed to be a better fit than any of the other Delphi companies (not a good fit, mind you, just a better fit than the others). Packard Hughes Interconnect, as it was named, had its headquarters in Irving, California, along with a printed circuit manufacturing operation. It had a components operation in Tijuana, a small operation in Arizona on a Native American reservation making wiring harnesses for the defense industry, and an operation in Foley, Alabama, making electronic components. A decision was made within Packard that Packard Hughes Interconnect needed to grow in order to be a viable asset to Packard.

Since Packard was in the automotive wiring harness business, the thought was that maybe Packard Hughes could pursue nonautomotive wiring business, and Packard could provide its expertise to help Packard Hughes become successful in this new business line. Hey, a wiring harness is a wiring harness, right?

At about this time, Boeing was looking to outsource some of its wiring harness business. It had been building wiring harnesses in the Seattle area with union labor, and they were having the same problems that Packard was experiencing with its union labor, i.e., too expensive and too inflexible. Because these were wiring harnesses for airplanes, the quality requirements were extraordinary and Boeing also was concerned about protecting technology, so they wanted to find a low-cost U.S. operation to which they could award the business.

When Packard Electric and Packard Hughes found out about this opportunity, they jumped at the chance. It didn't matter that no one in either company knew the first thing about aviation wiring. We knew a lot about automotive wiring and that was probably more than sufficient—you bet.

We expressed our interest to Boeing and they responded by giving us a set of prints for a 737 airplane to quote.

Well, we really wanted this business, so we were going to give them our best quote using our best technology (for automotive wiring, that is). The feedback I heard later, after I agreed (rather foolishly in retrospect) to become the manager of Packard Hughes' Alabama Operations, is that the quote contained automatic cutting and terminating, machine taping, and a few other tricks that can be used on automotive wiring, but, unfortunately, as Packard learned too late, could not be used on aviation wiring.

To make a long (and very sad) story short, Packard Hughes won the business, then went about finding some experts in aviation wiring that could be hired to help start up the new business and run it. (It would have been really nice if Packard had decided to seek the help of an expert before the quote was submitted.) The good news is that Packard had $20 million of new business. The bad news is that Packard Hughes experienced a loss on this business of about $10 million per year.

After this business had been won and was up and running, a decision was made that the guy running the Boeing business at Packard Hughes had to go. How could anyone lose $10 million per year on a $20 million package? (It isn't very difficult if the bid is half of what it should be, but that's beside the point.) I was asked to take on the job because I had a pretty good track record of fixing things that were broken. However, there was no fixing this disaster. We did win a second package of business from Boeing on which we did make $2 or $3 million, but that doesn't begin to cover a $10 million loss, and Boeing was not going to budge on the pricing of the initial package. (Believe me, we tried everything we could think of.) Sometime after I returned to the Mexico West Operations headquartered out of El Paso, the Boeing Plant in Alabama was boarded up and Boeing pulled their business back.

I'm sure Boeing was wiser for the experience. I'm not sure about Packard, based on its track record.

Let's Sum Things Up

Even though there are some gargantuan mistakes detailed in this chapter, I do not want to leave the impression that Packard never did anything right. For example, I was finally able to convince Packard to establish a Components Methods Lab, similar in concept to the final assembly preplanning Methods Lab, except that this lab would concentrate strictly on components. With their fine work, the problems constantly associated with the introduction of new components virtually disappeared. No longer were there any design problems that made the components difficult to use; no longer were there any mating problems; no longer were there problems plugging into the new components; no longer were there any problems with ringing-out and inspecting the components. In short, with the introduction of the Components Methods Lab, the introduction of new components ceased to be a problem.

Packard also continued to do a good job of identifying business opportunities around the globe and getting a good foothold in these new markets.

However, the mistakes detailed in this chapter were very detrimental to Packard's future (and some continue to be) and these mistakes just should not have been made. This chapter demonstrates several things in spades.

- No new process should ever be introduced without sufficient study, analysis, and debugging so that a company can be completely confident that the projected savings, in fact, will be realized and that the process will run as planned.
- Don't make the assumption that because a concept is old, it is no longer relevant. It may need to be updated, but it may still be the very best concept for achieving the desired results.
- Don't assume that an academic, with a background in business completely different than your business, has all of the answers. Every recommendation from a supposed expert must be looked at with a great deal of scrutiny to determine if it is valid for your operations.
- Don't pretend to be an expert or have knowledge in an area where you clearly don't. You may wind up making a decision that will come back to bite you.
- Unless you believe you as an upper-level executive have done a very poor job of hiring and training your lower level executives, it is a good idea to listen to them. Chances are they know a lot more about the current business than you do.

Even if Packard had not made any of the mistakes mentioned in this chapter, it is probable that Delphi Corporation would still have gone through bankruptcy. From all reports, the other companies within Delphi Corporation made even more blunders than Packard, all the while not being anywhere nearly as aggressive in moving business to low-cost manufacturing locations as Packard had been.

There is just no way that the Delphi Corporation could ever compete with automotive suppliers who were not saddled with totally noncompetitive compensation packages for union workers (compensation packages for salary employees have never been out of line with U.S. industry norms). However, Packard alone lost more than enough money with these very avoidable debacles to be able to fund the benefits for all of Delphi's salary retirees for eternity (benefits that have been terminated after they had been earned over entire working careers).

Chapter 4

Let's Get Lean ... Not

One of the things I stated several times in the last chapter, which was a review of some of the self-inflicted disasters created by Packard upper management over a relatively short period of time, is: This is something that Toyota would have never done. Starting sometime in the early 1980s, Toyota showed up on everyone's radar screen because of the tremendous inroads it was making in the worldwide automotive market, and, especially, the tremendous success it was having with its vehicles in the United States. GM couldn't seem to do anything right—from the poor quality of its cars to the poor decisions it made regarding models and platforms (e.g., the look-alike cars from all GM divisions under Roger Smith) to the poor performance of its plants.

GM finally decided that if you can't beat them, you might as well join them, and so it did. In 1984, New United Motor Manufacturing, Inc. (NUMMI) was formed as a joint venture between Toyota and GM in Freemont, California, which had been the home of Firebird and Camaro production in the United States before the plant was shuttered in 1982 due to poor performance. I suspect that the main reason Toyota was interested in such an arrangement was to get some experience setting up a manufacturing operation in this country before going out on its own. (I doubt if Toyota thought there was much, if anything, it could learn from GM about car manufacturing, and it was probably right.)

Toyota obviously learned its lesson well, as it now has wholly owned operations in Kentucky, Indiana, Alabama, Texas, and Mississippi as well as in Canada and Mexico. The company even agreed to allow the UAW to represent the NUMMI plant, which is something it certainly would not consider in its own plants. I often wondered if Toyota agreed to this knowing that some of the things it did to be successful would not be able to be done in a union environment, thereby denying GM full benefit of potential knowledge transfer.

GM obviously went into the deal with the idea that it could learn a lot from Toyota about how to run an assembly plant, especially one making small cars. The Saturn plant was started in 1990 in Springhill, Tennessee, incorporating what had been learned, and it probably was one of the more productive GM plants early on. However, I suspect this may have been a lot more a function of the new labor force in a new southern location than anything else. Whatever the case, the new plant with the new concepts and the Saturn car did not turn out to be the "Japan beater" that GM had promised. This plant no longer makes Saturn (a name plate which has since been terminated because a buyer could not be found by GM), and the plant was even considered for shutdown as part of a GM restructuring in 2007, until incentives by the state of Tennessee convinced GM to keep it open.

You can't blame all of the problems of Saturn on the assembly plant; this is certainly not the case. However, it also is clear that this plant was certainly no Toyota-like plant either. I'm sure a large part of this still had to do with the UAW, whose leadership is so short sighted it is incapable of seeing what's a few miles (or a few years) down the road (unless, of course, it can see it but doesn't care, because the current culprits will be gone when it happens). But, it's still pretty clear that either Toyota is a bad teacher or GM is a lousy student or it just may be a combination of both.

Toyota seems to be very open about what it does and how it does it. They have allowed countless competitors and noncompetitors alike to tour their facilities, and they appear to be very free with information and the tools they are using. But, my question is, do they intentionally mislead their visitors by what they present to them or in the way in which it is presented? Maybe, maybe not, but, one thing is for sure, some individuals (maybe more than some) come away with erroneous ideas of what makes Toyota such a great manufacturing company.

GM, for one, apparently had a very difficult time "seeing the forest for the trees," and the "trees" were all of the tools and practices that were observed in Toyota plants. And, what about the "forest" that GM didn't see (or understand)? They are the foundational principles and core beliefs of Toyota that were developed over five decades that dictate how its business is conducted. Much of this is not even written down and cannot be articulated because it is so ingrained in the fabric of tenured Toyota employees and which is intensively taught to all new employees.

Stephen Spear and Kent Brown, who I referred to in Chapter 1, spent four years studying Toyota, and in their white paper, entitled "Decoding the DNA of the Toyota Production System," (*Harvard Business Review*, Sept.–Oct. 1999) did a better job of capturing and articulating what truly makes Toyota tick than anyone else I have read. They stated, "Toyota does not consider any of the tools or practices—such as kanbans, andon cords, or *quality circles*, which so many outsiders have observed and copied—as fundamental to the Toyota Production System [TPS]. Toyota uses them merely as temporary responses to specific problems that will serve until a better approach is found or conditions change."

This is critical to understand. Toyota uses the various tools that have become synonymous with the TPS as "means to an end." GM and Packard used them as "ends in themselves." This is a tremendous distinction and, I believe, the main reason for the tremendous differences in results in what has been achieved by Toyota on one hand and GM and Packard on the other, as well as just about every other company that has tried to copy what Toyota has done.

What I will try to demonstrate in this chapter, through several examples, is that Packard was a lot closer to using the TPS in 1971, when I first started at Packard and before anyone knew much about Toyota, than it was in the twenty-first century after a couple of decades of trying to copy Toyota, but not having a clue as to what to copy.

Just What Is the TPS?

Toyota Automatic Loom Works, Ltd. was established in 1926 by Sakichi Toyoda, who was the inventor of a series of manual and machine-powered looms. In 1933, an automotive department was established and run by Toyoda's eldest son, Kiichiro. In 1936, many decades after GM built its first car, the first Toyota car was built and, in 1937, Toyota Motor Company was spun off under the leadership of the younger Toyoda.

The Japanese auto market was very small in relation to the U.S. market in the early days, and their manufacturing techniques were quite primitive. In the 1950s, Toyota sent a delegation to the United States, led by the engineer Taiichi Ohno, to study the U.S. auto industry, which was by far the world's largest and most advanced, as well as to study some of the other industries in this country, in an attempt to gain insight on concepts they might be able to employ to improve their operations.

They visited several Ford plants in Michigan and, on the whole, they were very impressed with what they observed; Toyota management would often say in later years that much of what they learned that enabled them to develop their manufacturing systems they learned from Henry Ford. However, there were some things with which they were less than impressed within the Ford factories: especially the large amounts of inventory that were observed onsite, the apparent unevenness of work being performed within the various departments on most days, and the large amount of rework at the end of the process.

After spending a lot of time at Ford, the delegation spent time at a Piggly Wiggly® supermarket, and here they saw something that impressed them greatly. They marveled at the just-in-time distribution system that was in place within the store and how the supermarket only reordered and restocked goods once they had been bought by customers. This inspired what would become the now-famous just-in-time (JIT) inventory system.

Ask most knowledgeable people what they know about the Toyota Production System and they will mention JIT, elimination of the seven forms of waste, and the various tools that have received notoriety, i.e., andon cords, kanbans, quality circles, kaizen continuous improvement boards, poka-yokes, etc. While all of these play some part in TPS, they are not what makes Toyota tick and not what makes them successful.

Spear and Bowen stated that, in their four-year study, they had observed many cases in which Toyota actually built up its inventory of materials as countermeasures against such things as unpredictable downtime or yields, time-consuming setups, and volatility in the mix and volume of customer demand. I'm sure this comes as a surprise to many who have been led to believe that Toyota does everything with next to no inventory, assuming it is not already being done with a single piece flow.

Remember, they also said that Toyota does not consider any of the tools or practices, which so many outsiders have observed and copied, as fundamental to the TPS. They are used by Toyota merely as temporary responses to specific problems. Even the elimination of the seven forms of waste is far from straightforward, as none of the forms of waste is independent or isolated from the other six. If you suboptimize and eliminate or greatly reduce one of the forms of waste, you may well be increasing waste in one or more of the other areas many times over. I will get into this in much more detail later in this chapter.

So, just what is the TPS? I think the best definition I can come up with is this: It is a well-founded, commonsense, manufacturing system based on unwavering core principles and beliefs that are taught, understood, and utilized throughout the organization. I believe that the reason Toyota has been so successful is that they just don't do a lot of stupid things. I'm sure that they make mistakes, but, based on their foundational beliefs, the mistakes they make have limited risk and, therefore, limited negative consequences. Contrast this to what you learned about Packard in the last chapter. Packard, by contrast, has done, and continues to do, lots of very stupid things, and the consequences are often far reaching and very expensive.

For the fun of it, let's review those core principles as described by Toyota in *The Toyota Way 2001* and see how Packard stacks up. After each principle, I will give Packard two grades. The first is how I would rate Packard in 1971 when I started and the second is what I would rate Packard 30 years later when I left the company, after they had spent more than two decades trying to become Toyota.

1. Base your management decisions on a long-term philosophy, even at the expense of short-term goals. (F → F)

   From what I have read and observed about Toyota, this has been practiced by them from the start. They obviously sacrificed some short-term profitability along the way to build the company to what it is now (which is one of the largest and most profitable companies in the world). But, was it worth it? I'm sure that GM, which should have gone through Chapter 11 bankruptcy and is still 26% government owned, and Delphi, which recently emerged from bankruptcy and is now a shadow of it former self, would give a resounding "yes" to that question, as would the financial institutions that were partly responsible for GM's short-term outlook.

2. Create continuous process flow to bring problems to the surface. (B+ → C–)

   The problem is that I don't think Packard management ever understood what this meant. I often got the impression that some executives at Packard thought it meant that, in the ideal system, every piece of material in the plant should be in constant motion, only stopping long enough to be processed before moving on to the next process. Of course, this is absurd, and Toyota doesn't do this. They use reasonable and justifiable inventories in front of processes, like most everyone else, although they do a better job of controlling inventories than most.

What I believe it means is that there should be well-defined, well-organized, orderly, and controlled flows of materials from receiving through the plant and then out the door, all based on customer demand. From the day I started at Packard, this was the case, at least until remote cutting and lead prep were introduced. There were some other bad decisions made regarding material flow in an effort to fix something that wasn't broken. A couple of costly examples will be given in this chapter.

3. Use pull systems to avoid overproduction. (B+ → C)

    One of the things I could never quite figure out was why so many Packard executives would continue to say that we needed to implement pull systems in order to get our inventories and production under control. The reason I could never figure it out was not because pull systems are not a good idea, they are. It's just that I had never known anything but a pull system to be used at Packard, at least not in the cutting, lead prep, or final assembly areas.

    The problem is that the pull systems we used did not involve kanban cards; they generally involved a reorder point on a computer program, an empty rack, an empty tub, or an empty location. These were very effective pull systems for our type of production, but they just didn't look like Toyota's kanban system.

    Over a period of several years, kanban systems were tried in several plants in cutting, lead prep, and final assembly areas, none of which was successful or which lasted very long. They generally resulted in a lot of lost cards and a lot of frustration; but you have got to give our management an "E" for effort. They just didn't want to give up until we had a "real pull system" in place.

4. Level out the workload. (A → C–)

    When I hired on at Packard, if Packard deserved an A+ in anything, it was this. I am confident that the preplanning process (described in Chapter 7) was unsurpassed (probably unequaled) in the world. Unfortunately, while I'm sure Packard will still tell you they preplan, it is not even close to the quality of the product that was achieved in the early days. By the way, the "suit" who was responsible for the IPS (Integrated Production System) and IPS II debacles described in the last chapter tried very hard to get engineering to drop preplanning in favor of standard data breakdowns. This would have been a disaster, as I will talk about in Chapter 7.

    In the early days, Packard, with the use of the excellent efficiency control system that was in place, also did a very good job of determining manpower requirements and properly allocating these resources throughout the harness-making operations. When this system went out the window, so did the ability to properly determine and allocate resources.

5. Build a culture of stopping to fix problems, to get quality right the first time. (B → C)

    The truth is that Toyota very rarely has to stop the line to fix problems. They use team leaders to help fix problems if operators need assistance, so that the line does not have to stop. Toyota surely understands that allowing a conveyor line to constantly stop, regardless of the reason, is a recipe for disaster.

    The strong manufacturing supervisors at Packard, in the early days, understood this extremely well, and, therefore, they used a similar philosophy to minimize the need to have unscheduled stoppages of the line. If an operator encountered a problem in a station, a universal operator, a repair operator, or a relief operator (I know that this is a lot of different job classifications, but you can thank the IUE and weak management for that) would be sent to help the line operator so that the problem could be fixed without stopping the line. If the line had to be stopped because of the severity of the problem, it would be done, but everyone knew it was a big deal, and stopping the line, unless there was a very good reason, was serious business. Of course, stopping any non-paced operation to fix a problem was standard operating procedures within Packard. If Packard had been a little more quality conscious in the early days, I would have given them an A. However, back in those days, you would often hear the refrain, "If in doubt, ship it out."

    I do believe that Packard became more quality conscious every year, and that is a good thing. What is not a good thing is that they allowed that thinking to unnecessarily impact productivity in a negative way. It became commonplace for lines to constantly shut down, especially when andon cords were introduced in a couple of plants (an andon cord allowed any operator on the line to shut it down, whereas before, a line could only be stopped by the first operator, who had to be asked to stop the line, and who kept a record of who did the asking). My observations were that little, if any, follow-up was made by the supervisors in Mexico when a line did shut down, which was certainly not the case in the old days in Warren. So, what message did that send to the operators? The message was that you could shut the line down, keep it down for as long as you needed to fix the problem (or longer), and then have it restarted without any repercussions. Can you see why this might have created some problems with productivity?

As if what I described above didn't create enough problems, another practice crept into many of the Mexican plants, which was even more damaging. Frequently, a defect on a wiring harness would not be discovered until the off-line electrical ring-out station at the end of the process. There was a concern that the operator responsible for the defect would not get proper feedback if the defect was repaired on the ring-out board, so the practice was to return the harness to the line for repair.

This might sound like a good idea, but only if you believe there was no other good means to provide as good as or even better feedback, and you do not consider the likely ramifications. When a harness was put back on the line, at a minimum, one harness of production would be lost, and, often more, because it would take longer than the established station cycle time (line speed) to put the harness back on an empty board on the line. Some of the routing pins on the boards also were stressed in the process of trying to return the harness to the board, which had a negative impact on future dimensional control. And, depending on the severity of the repair, one of two things would happen: (1) on easy repairs, the offending operator would finish the repair in a few seconds and then take a break for the remainder of the station cycle time, or (2) on complex repairs, it would take longer to make the repair than the station cycle time, so the line had to be stopped again. Of course, while all of this was going on, all of the other operators on the line would be receiving an unscheduled and undeserved break.

As you know, I am a big believer that you will get the performance that you motivate (or, another way of saying it is: "If you reward undesirable behavior, you will get more of it; if you punish undesirable behavior you will get less of it."). Making a defect is a bad thing of which you want to get less. What do you think the likely outcome would be if an operator's punishment for making a defect was to receive a nice break for himself and his fellow operators? I'll give you a hint. It's kind of like not doing your homework and being punished by being forced to go to the movies (after being given money for both the show and popcorn).

What do you think the likely outcome would be if your punishment for making a defect was not a break, but instead a friendly visit from the supervisor asking what he can do to help in order to avoid future defects? See, this isn't really very tough, when you think about it. I could just never figure out why Packard kept making such errors in judgment.

6. Standardized tasks are the foundation for continuous improvement and employee empowerment. (A → D)

In the early days, Packard was exceptionally strong in the area of standardized tasks, because every station on every line came out of the Methods Lab with very detailed methods and quality considerations. Excellent methods work also was done on all off-line operations (stationary final assembly build stations, subassembly stations, nonconventional splice stations, nonconventional lead prep operations, etc.) to include detailed left hand–right hand build methods, detailed layouts including the positioning of all materials, hourly standards, and quality requirements.

Nor was there any pressure or encouragement to be different for difference sake. One Packard assembly plant would look just like the next. If someone came up with a demonstrably better idea (meaning that it proved to save money, improve quality, improve safety, etc.), the idea would be adopted throughout Packard, generally by the Methods Lab and Industrial Engineering implementing the idea on all future production processes (there was only one Methods Lab for all of the Warren plants). Although I don't remember seeing it written down anywhere, there was a widely held understanding that standardization was very important to establish a foundation for effective manufacturing and continuous improvement.

I have already talked briefly about the degradation of the preplanning done in later years in Mexico, compared to how it had been done in the early years in Warren, especially for all of the off-line operations. However, even with this slippage, the preplanning done in Mexico was still much better than average for industry, because many (if not most) companies don't do anything resembling preplanning at all.

The biggest problem in Mexico regarding this Toyota principle was (is) in the understanding of and implementation of the concept of continuous improvement. Toyota very much understands the importance of continuity, consistency, and standardization, and its pursuit of continuous improvement is done in such a way as to not negatively impact these things. In recent years, Packard's concept of continuous improvement made it impossible to achieve these critical factors.

Through the words and actions of Packard executives, manufacturing supervision came to the clear understanding that continuous improvement meant continuous change. If a plant manager's plant looked just like every other plant, he was judged as not being innovative and creative by the honchos (those guys/gals who controlled his future), and they made it clear that this plant manager obviously didn't understand the concept of continuous improvement (as understood incorrectly by Packard executives).

Tremendous importance began to be placed on innovation and creativity, not how productive a plant was being run, just whether or not new and exciting things were being done within it. When wide-aisle tours were conducted for Packard executives, care was taken by manufacturing supervision to point out all of the new innovations, because that was what the execs wanted to see. Those who could show a lot of new stuff, regardless of whether or not there was any tangible benefit from it, got a lot of "atta-boys." Those who didn't were clearly left with the impression that they weren't getting the job done. Who cares about how productive a plant is, let's see some new stuff. How misdirected can you be?

As you can imagine, things got to be nuts. Every plant started looking different and was doing different things, and everyone was claiming that they had the best ideas; although no one ever had any facts, just opinions. All the while the performance of the plants continued to slip. There are only so many things on which anyone can focus at one time and do them effectively, so you had better be focused on the most important thing.

7. Use visual controls so that no problems are hidden. (B → C)

Throughout its history, Packard has done a pretty good job of using necessary visual controls. There were two visual controls used at Packard in the early days that were very helpful, but which were eventually scrapped. One was a Quality Board at the end of every conveyor line, which was in the direct line of sight of the supervisor, on which the end-of-line inspector recorded every defect made on the conveyor by type and by station. If you think this did not have a positive impact on the quality coming off of the line, you are very mistaken. And, it gave the supervisor a good tool to monitor in order to provide proper follow-up. The other was the elimination or curtailing of the line startup performance charts. These were also great tools to help understand just what was going on as the lines were ramping up (from both a quality and productivity standpoint) and where attention needed to be focused.

Other than these two glaring omissions, in recent years, if there is a complaint to be made, it is that Packard used too many visual aids. There were so many visual aids in some of the information centers that it was a real chore to keep them up to date (and frequently they were not), but no matter, they were rarely observed or acted upon anyway.

8. Use only reliable, thoroughly tested technology that serves your people and processes. (D– → D–)

    Some of you are probably thinking, based on some of the examples from the last chapter, that I am being too generous with this rating. I'm giving credit for the fact that many of the processes Packard has used over the years have been thoroughly tested and they are reliable. But, if your point is that this was far from a core principle of Packard, I can't put up much of an argument.

9. Grow leaders who thoroughly understand the work, live the philosophy, and teach it to others. (B → D–)

    In the early years, there was clearly an effort to prepare good leaders through real-life experiences in engineering and manufacturing, and, at least in Industrial and Methods Engineering, some outstanding training was given. There was also a good deal of teaching done through example and osmosis.

    This is just no longer the case as the once strong training programs have largely gone by the wayside and beneficial real-world experience is sorely lacking. When I returned from South Korea, I found out that the once excellent IE Training Program was nonexistent, not just stale due to a lack of updating and enhancement. While overseas, I compiled and wrote an IE Manual complete with a detailed training program (see Chapter 2), and I tried very hard to have it institutionalized. There were some successes, but intensive IE training never again became imbedded in the fabric of the company, much to its detriment.

10. Develop exceptional people and teams who follow your company's philosophy. (B– → D–)

    The obvious problem here is just what is the company's philosophy that the exceptional people and teams are supposed to be following? Are they the sweet sounding, flowery words that are contained in the Packard Principles; in the Excellence Concept; and in Packard's Vision, Mission, Objectives, and Strategies Pamphlet? Or is it what Packard demonstrated through its actions?

    Packard has always been very reactionary to the three-month P&L statement, mainly because its parent, GM, lived and died by the three-month P&Ls. And, GM would give dictates to its subsidiary companies, as it felt necessary, to get the results it believed the shareholders, financial institutions, and markets demanded. Virtually all of these dictates came in the form of demands for reductions in inventory and reductions in headcount, mainly, I think because these were the two numbers that the executives could understand and get their arms around.

I'm sure most of you reading this book are aware that Japanese companies are known for providing a virtual lifetime guarantee of employment (although, my experience is that they are able to get rid of "dead wood" by shaming them and getting them to resign). GM and Packard made no such guarantees, especially in the Mexican Operations. I certainly understand and agree that there must be a way to separate poor performing employees from a company if that company is to remain healthy. Sometimes it is even necessary to cut deeply enough that good performers are impacted if a company is to stay afloat.

However, neither of these two situations existed within the Mexican Operations on more than one occasion when some outstanding people were sent packing. It was simply a matter of the projected three-month P&L not looking as good as GM wanted, due to a typical blip in business, and, because of this, GM would issue headcount reduction edicts (for salary workers). These edicts never seemed to be questioned by top Packard executives, even though we all knew we were going to need to hire new employees within six months, probably less qualified and certainly less experienced employees than the ones we just booted out the door.

I'm sure that it cost us more than we saved every time we had to dismiss a good employee, not to mention the negative impact on the overall attitudes and level of dedication of the workforce. Yes, "Packard recognizes the contributions of every individual and understands that its employees are its most valuable resource and will be treated with dignity and respect," just like the Packard Principles say.

There is no question that exceptional people are going to be a lot harder for Packard to find in the future than in the past, at least in North America.

11. Respect your extended network of partners and suppliers by challenging them and helping them improve. (D → D)

What I know about Packard as a supplier is that they were not great when I was in South Korea, and I know because we were ordering a sizeable number of part numbers from Warren. I designed and had programmed a Materials Management System that, among other things, provided a supplier rating for every supplier based on their adherence to ship dates, quantities ordered versus quantities shipped, and the quality of their materials as received in our warehouse. Packard's ratings were *about middle of the road, and we were comparing them to suppliers throughout South Korea, Asia, Europe, and the United States. We even sent a copy of the automatically generated reports monthly to each of our suppliers with their specific scores, as well as where they were positioned on a scatter chart, which showed the performance of all our suppliers (although we left off the company names). Most of our suppliers, who were not in the top 5%, contacted us and let us know that they appreciated the information and intended to improve. We never heard from Packard.*

Regarding Packard as a customer, you would think, after reading the Packard Principles and other Packard documents, that Packard would be a fabulous customer (unlike what you read in Chapter 1 about GM's reputation as a customer). There is a Packard Principle entitled "Value of Suppliers," another entitled "Responsibility to Society," another entitled "Trust in Relationships," and yet another entitled "Integrity." They all say about what you would expect them to say; Packard just seemed to have a hard time living up to its lofty ideals. Based on everything I heard first hand from our materials people, Packard was a lousy customer. Seems we had a problem paying bills on time and liked to browbeat suppliers, just to name a couple of small problems.

From the feedback I received from my materials friends, it sounds like Packard was strong on "challenging" its suppliers to improve, but a little weak on the part about helping them, unless you consider making dictates helpful.

12. Go and see for yourself to thoroughly understand the situation. (B → D)

    As I explained earlier in this book, in the early days, the bulk of the executives within the various operations areas had a background that included experience at the lower levels in manufacturing and/or engineering, and they knew the business well. As a result, they were comfortable in the plant and knew what they were observing. We still had our share of wide-aisle tours and dog-and-pony shows, and I would have liked to have seen some of the plant managers and superintendents, as well as engineering managers and superintendents, in the plant more often. However, in total, the operations executives did get into the plant for significant events and were not making decisions with a glaring lack of knowledge and understanding.

    As time wore on within Packard's North American Operations, the quality time spent in the plants by operations executives seemed to become more infrequent, although there never seemed to be any shortage of the nonquality time, i.e., wide-aisle tours and dog-and-pony shows. If more quality time had been spent in the plants by these folks, I'm confident that many of the problems I discussed in the last chapter and will discuss in this one would have been much less severe and long lasting.

Based on my experiences as a manufacturing foreman in Warren and my experiences having manufacturing responsibility in South Korea, Alabama, and to some degree in Mexico, the concept of "managing by walking around" is essential and by far the most effective concept to improve manufacturing performance. Of course, knowing what you are doing while walking around is also important, but just being consistently on the floor is going to get you 75% of the way to where you need to be.

13. Make decisions slowly by consensus, thoroughly considering all options; implement decisions rapidly. (D → D)

    The last part about implementing decisions rapidly once they have been made has never been a big problem at Packard, and I would give them a B. The problem is the first part, on which Packard would rate a D– at best, both in the early days and 30 years later. I sat in a lot of meetings over the years with high-level executives where input was requested, but I rarely got the impression that it was anything more than an effort to appear open-minded. I never saw much evidence that they ever were. Often, decisions seemed to be rushed through, and certainly without considering all of the possible fallout.

    I was taught as a young Industrial Engineer that all plausible options needed to be developed, analyzed, and evaluated by use of scientific methods when trying to solve a problem, improve a process, or initiate a new production plan. This was religiously done within IE for several years, and a few of us tried to maintain this discipline, but, by and large, it was not something that was done at the higher decision-making levels of the company. Many of the big blunders you have read about, and will yet read about, attest to this.

14. Become a learning organization through relentless reflection and continuous improvement. (D → D–)

    Relentless reflection has never been a strong suit at Packard. If it had been, there would not have been so many stupid new blunders, many of which looked surprisingly like stupid old blunders that had already been made. Once the first semiautomated line failed miserably, serious reflection and analysis would have avoided future failures of the same type. Once it became evident that the decision to terminate the productivity control system was resulting in significantly decreased performance, the decision would have been made to reinstate it (preferably with updated technology). Packard was very poor at learning from history. Packard not only failed to learn anything from the mistakes of others (unless it ended up copying others' mistakes while benchmarking), it didn't learn much from its own mistakes.

In the early days, there was some focus on maintaining and improving the basics, which is really what Toyota is talking about. This was lost over the years as Packard kept trying to find the silver bullet that was going to fundamentally transform the company. Continuous improvement at Packard came to mean doing something significantly different than what was currently being done (i.e., a different system, process, practice, method, tool, layout, rack, etc.) to try to generate improvement. This propensity only tended to make things worse as focus was removed from the basics, which were absolutely critical to success.

Instead, what Toyota means, at least as I understand it, is to make incremental improvements to what is currently being done by identifying opportunity areas and implementing enhancements, while never backsliding in any of the other areas. (This is not to say that Toyota would not consider big changes in what they are doing, just that they would not implement one of these changes unless it was thoroughly analyzed, justified, and debugged before going into production, and then they would make it the standard throughout their operations.)

To make a sport's analogy, let's assume that during the course of the year our quarterback is getting sacked too often. Toyota's approach would be to do a careful study. Let's assume the study established that the left tackle was using poor pass protection technique. Their solution would be to provide the training and support necessary to help the left tackle achieve the level of performance desired and needed by the team. Packard would say, "Wow, our quarterback is getting sacked a lot. Let's go get us another offensive line (or maybe another quarterback)." Who do you think is going to get more for the money?

I suspect that if Toyota was being totally honest and did an evaluation within each of its facilities on all 14 of these principles, it would find that there was some room for improvement within each facility, but I doubt that any of the facilities would rate less than a solid B on any of the core principles; these principles are just too deeply engrained in the souls of Toyota employees.

So, let's do a little review of how well Packard did.

Let's start with Packard back in the early 1970s. Being as objective as possible, I gave Packard two As, two B+s, five Bs, three Ds, one D−, and one F. This is obviously a good news–bad news story. They scored very well on 9 of the 14 Toyota principles (before anyone at Packard knew that these principles existed), but they scored very poorly on 5. Overall, I think you would have to give Old Packard a passing grade, especially since it didn't even know that it was taking a test.

But, what about the New Packard after not only knowing the test, but trying to become Toyota for over two decades. You would think that Packard would be able to marginally improve the high marks, or at least maintain them, while greatly improving the low ones. But, three Cs, two C–s, four Ds, four D–s, and one F fall just a tad short. You explain it, I can't (unless, of course, you consider the possibility of incompetent management).

Peter Drucker said that the productivity of work is not the responsibility of the worker, but of the manager. W. Edwards Deming told Ford that management actions were responsible for 85% of all problems in developing better cars. I think they were both being very reserved in their comments. I think Packard management was closer to 99% responsible for the failures of Packard to achieve good scores on these 14 TPS core principles.

Toyota Production System Rules

Before I go on to a few of Packard's feeble efforts to implement new processes or concepts it believed to be in compliance with the Toyota Production System or Lean Concepts, I want to list "The Four Rules of the Toyota Production System" that work in harmony with the Toyota Core Principles stated above, which were discovered by Spear and Bowen during their four-year study. These rules help provide a little more insight into what Toyota has done, and continues to do, to be successful, and you will see that they are in total harmony with the 14 principles.

> Rule 1: All work shall be highly specified as to content, sequence, timing, and outcome.
>
> *This is extremely important in a manufacturing operation and it is directly supported by principles 4, 6, 7, and 8 and indirectly supported by others. Old Packard did an excellent job following this rule; New Packard, not so much.*
>
> Rule 2: Every customer–supplier connection must be direct, and there must be an unambiguous yes or no way to send requests and receive responses.

This rule is also very important and is directly supported by principles 2 and 3, and indirectly supported by others. Packard has always done a pretty good job following this rule; its just that many Packard executives didn't realize it because they were looking for a kanban.

Rule 3: The pathway for every product and service must be simple and direct.

Again, this is an important rule supported by principle 2 and others, and one which Packard understood very well. A great deal of effort was made to ensure effective layouts and smooth material flow within the plants. New Packard got a little bit carried away on the concept of "vista," which I will talk about later, but in total, Packard has certainly understood the importance of this rule, at least until the implementation of remote cutting and lead prep.

Rule 4: Any improvement must be made in accordance with the Scientific Method, under the guidance of a teacher, at the lowest possible level in the organization.

All of the rules are equally important, but this one just might be "the most equal." It is supported directly by Toyota principles 9, 10, 13, and 14 and indirectly by virtually all of the others. At one time, Packard understood this rule, taught this rule, and used this rule when making any changes or improvements in the plant (at least, that was the case within Industrial Engineering).

Oh, how things have changed at Packard. There is another saying we used in IE that is very appropriate here. It goes like this: "Without data you are just another person with an opinion ... unless, of course, you are at a level where your opinion becomes data." And, believe me, there were a lot of executives at Packard with opinions.

I don't think the importance of this rule can be overstated. If you review all of the gross mistakes that Packard committed in the last chapter, and what I will write about in the rest of this chapter, you will see that every one of them could have been avoided by adherence to this rule. These awful decisions arose because of top-down dictates from executives who didn't have all of the necessary data to make intelligent decisions, and these decisions were not questioned (at least, not by the right people with the right degree of forcefulness).

There were certainly employees at Packard that knew, beyond any shadow of doubt, that these decisions were bad, but they were generally employees who were not at the highest levels in the company. It is necessary for an executive to have a vision, but it must be tempered with reality, especially in regards to the implementation of process change. Reality exists on the shop floor. This is why Toyota believes it is critical to make decisions regarding improvements at the lower levels of the organization, based on careful analysis through utilization of scientific methods. Toyota just would not have made the glaring mistakes that we have talked about and will talk about in the rest of this book.

Packard Develops the PPS

Packard so thoroughly worshipped Toyota in the early 1990s that the Packard Production System (PPS) was developed, which, as you might surmise, was almost a carbon copy of the Toyota Production System—at least what Packard thought the Toyota Production System was. Based on how well Packard was able to achieve the Toyota Principles shown above, if you guessed that within the PPS there was a lot of misunderstandings, omissions, and improper additions to the TPS, you would be correct. Throughout the rest of this chapter, I will give a few of the more outlandish attempts by Packard to become Lean.

Again, I am revealing these blunders, not to embarrass anyone, but so that those of you reading this book can learn from the mistakes of Packard, which, at this time, was the world's largest supplier of wiring harnesses and components to the automotive market. Remember, a really wise person or organization is one that learns from the mistakes (and experiences) of others. It is too costly and painful to have to learn everything from your own mistakes.

What We Need Is Vista

I'm not sure where the concept of "vista" originated within the Mexican Operations, whether it was something one of the Packard executives read about or just dreamed up. Of one thing I am sure, it was an exceedingly expensive and ill-advised attempt to accomplish a Lean objective, but it was anything but Lean. Apparently, the idea behind vista was to create a manufacturing environment that would be very conducive to good flow, would facilitate pull systems, would enhance a supervisor's ability to control processes, and would create a better and safer working environment for the workers.

While all of these things sound like very noble objectives, one must always ask the questions: What are the actual savings and tangible improvements we can expect to achieve and what is it going to cost to implement it?

Had these questions been asked, evaluated, and then answered, there is no possibility that vista would have gotten the green light.

There were a couple of aspects to vista as envisioned by Packard management. Part one was that everything on the plant floor should be kept to eye level or below, if possible. Part two was that sufficient space should be allowed around the various processes so that the smooth flow of materials would never be inhibited. Again, these sound like good objectives, until you get into the details.

When I was still in South Korea, I got a call from the engineer in charge of the plant layouts within the Mexican Operations. He was calling to see if I could help explain why the European Operations were able to produce twice as many standard hours of production, within a given amount of floor space, as Packard was within the Mexican Operations. Apparently, Packard Warren executives had started asking questions since they had become aware of the gigantic discrepancy as pointed out to them by Packard's European executives. Because I had been in the European Operations, and also intimately knew the Mexican operations, he thought I could probably help explain it.

I told him that probably a third of the difference was due to a higher level of productivity and probably another 10 to 15% was due to the type of racks that were used. In Europe, and also in the South Korean Joint Venture, we used racks on the conveyor lines that not only held materials for immediate use by the operators, but there was also some online storage capacity for final assembly materials. I told him that I couldn't explain the remaining 50% variance, but that I would be returning to the states on home leave in a few weeks and I would make arrangements to get into the Mexican plants to take a look.

Once I entered the Mexican plants, it became obvious very quickly where the other 50% was. There were open areas between each set of adjacent conveyor lines wide enough to hold a bowling alley. The standard center-line to center-line spacing of conveyors had always been around 21 feet. This would give ample space to get a service buggy between the racks on adjacent lines; just not enough space to move two service buggies between lines such that they could pass side by side with no interference, even though this was seldom necessary and not that big an inconvenience when it was. The new spacing was closer to 28 to 30 feet. Sure, it looked nice and open and decongested, but at what cost?

The second part of vista required that all of the high racks on the lines, which were used to hold the longer wires, be eliminated. The way Packard dealt with this challenge was to coil the long wires in the lead prep area before bringing them to the final assembly area and servicing them on the line. Prior to vista, the standard method of handling long wires (unless they had tangling terminals on both ends, which would require coiling) was for the final assembly service person to take a bundle of long wires and put the bundle of wires directly onto the line (on one of several different types of specially designed racks), then properly comb out the wires per method such that the operator could pull an individual wire from the rack with minimum difficulty. Occasionally, the long wires would become entangled and the service person would have to restraighten them, but for the most part, the problems were minimal.

With the implementation of vista and coiling, the new method added the following nonvalue-added work: having a lead prep operator coil each wire and put a piece of tape on each coil to control it, having service personnel move the coiled leads to the line and place them on a rack, and having the conveyor operator remove the tear tape and uncoil the lead before building it to the board. This added a significant amount of labor to the process, not to mention the cost of the discarded tape.

One other thing that was done at about this time, which was done at least in part to satisfy vista, was to remove all of the splice presses that were integrated directly into the conveyor lines and place them either in the lead prep area or the area at the back end of the final assembly department. This added a significant amount of servicing and material handling labor, and also added significantly to floor space requirements.

All of these things accounted for the other 50% discrepancy in floor space utilization, which I couldn't explain before I toured the plants because I had no idea what had been done. It is bad enough to add significant nonvalue-added labor to the plant, but to throw away valuable floor space for no good reason, other than plant beautification, makes no sense, especially when you are supposed to be squeezing nickels.

Benchmarking: To Do or Not to Do

One of the tools that was closely associated with Lean and the TPS within Packard (and, if not an actual element of Lean or TPS, was at least considered a first cousin) was the concept of benchmarking. Benchmarking is defined in The Free Dictionary.com as "to measure (a rival's product) according to specified standards in order to compare it with and improve one's own product."

From the very definition, it is obvious that benchmarking is far from an exact science. While some companies appear to be open and sharing with their processes and procedures (for example, Toyota, at least on the surface), every company is in business to be profitable and to stay in business. Giving away trade secrets, or sharing concepts that give a company a competitive advantage, is probably not the best way to accomplish these two main goals.

It is also not that easy to establish just what constitutes a best practice. It must be considered in the context of everything else that interrelates to it within the company. For example, the Green Bay Packer power sweep, under the tutelage of Vince Lombardi, was considered a best practice in professional football, but many other teams tried to duplicate this best practice without success. Why? Because they did not have the other ingredients on their team that Green Bay possessed that allowed them to be successful. Taking a single concept out of an integrated system and trying to implement it somewhere else, without benefit of the rest of the support system, is not likely to lead to great success. It must be noted as well that some concepts are just not transferable. What might be a best practice in the automotive wiring industry might be totally inapplicable to the aviation wiring industry, as we shall see later.

Another problem with benchmarking is that it is very easy to be misled, even when it is not done intentionally. Benchmarking is not always done in a particularly thorough and in-depth manner; it's more like how wide-aisle tours and dog-and-pony shows are conducted in most facilities. Even friends from other companies or organizations, and certainly other groups within your own company, are going to do their best to present what they have done in the best possible light, with all of the beautiful charts and graphs they can muster. They may not intentionally try to mislead others about the benefits of what they are presenting versus the costs, risks, and tradeoffs, but it is the nature of the beast that this is what will often happen. When involved in the act of benchmarking, the best advice is "buyer beware."

I have made this brief review about benchmarking because the next three examples of rather regrettable mistakes made by Packard, in an effort to become Leaner, were made, to some degree, due to benchmarking activities that had been done by Packard executives.

The message here is not to steer clear of the concept of benchmarking; certainly some good can and does come from it. The message is to use caution and to make sure that a thorough and complete analysis is made prior to implementing a concept that is benchmarked, and, if a decision is made to proceed, do it in such a manner that risk is minimized (e.g., in a pilot program). Do not make the mistake that Packard seemed to make repeatedly, i.e., observe or find out about something someone else is doing; assume that because they are doing it, it must be a good idea; assume that it must be transferable to your operation, even though your businesses may be quite different; and then implement the concept in a big way. This is certainly a recipe for disaster.

Before I go on to Packard examples, I'll give an example of benchmarking that was done using a Packard concept. After I left Packard, I consulted for several years for another large automotive components supplier, and one of their product lines was automotive wiring, although they were small in comparison to Packard. But, they wanted to dramatically grow this line of business. Their assumption was that since Packard was the biggest and the oldest wiring harness maker, they must have had everything figured out. What they didn't understand is that sometimes companies are big (and sometimes, even very successful) in spite of, and not because of, certain of their processes, practices, or concepts.

This was the case regarding one concept that this company benchmarked from Packard. This company had historically used plywood in the making of final assembly build fixtures (boards). They became aware that Packard had switched to perforated steel plate a few years before I started consulting with them so they figured that perforated steel plate must be the right answer. They proceeded to convert all of their boards over to perforated steel plate whenever it was necessary to build new tools for model change or the introduction of new business.

The problem is that there were many within Packard (me included) who believed that a very bad decision had been made to go to perforated steel plate in the first place. The two main reasons that Packard decided to go to steel plate were: (1) steel plate is presumably reusable and (2) coordinates could be put on the boards that would make duplication easier and eliminate the need for a wire template to ensure the dimensional integrity of each board (and, oh yes, I almost forgot, steel also looked sexier). However, even in these two areas, there was little, if any, benefit to using perforated steel plate boards over plywood. Sheets of plywood also could be reused and had a life expectancy about equivalent to that of perforated steel in practice, and it was relatively easy to make exact duplicates of boards made with plywood, so long as the quality of the fixtures put on the boards was controlled. The main advantages of plywood boards are that they are a small fraction of the cost and weight of the steel boards, and they are also more precise, because the holes in perforated steel have a center distance of about a quarter inch.

I helped this company in their operations in the Philippines for several months, and one of the things I convinced them to do was to go back to plywood boards, albeit, with completely redesigned board fixtures that were standardized for dimensional control and that were much more functional. The result was that they were able to build several new lines of boards, with excellent dimensional integrity and without the use of a template, at well under half of the cost of perforated steel boards.

The moral of this story is: "Be very careful what you benchmark; their idea may not be as good as yours."

Kitting Looks Like a Cool Thing to Do

One of the directors we had in the Mexico West Operations had actually spent time at the start of his career as an hourly worker and later as a foreman, but he was so focused on his career that I believe it would be fair to categorize him as a "suit." It was clear from the time I started working closely with him that his desires to reach the highest levels of the company were paramount, and he would do anything to get the recognition that might help him achieve his goals.

A big focus of Packard in the early 1990s was innovation and creativity. As I have pointed out, this focus often led to some very unpleasant experiences, but that didn't seem to dampen enthusiasm within the company for these two characteristics (innovation and creativity) that were perceived to be very valuable. This director, through some form of benchmarking, had somehow come to the conclusion that kitting materials for the final assembly lines (the process in which individually separate but related items are grouped, packaged, and supplied together as one unit), which had never been done at Packard (and for very good reasons), was what was needed for the startup of two plants in Victoria, Mexico, referred to earlier. In his mind, this was something innovative and creative, and, when it succeeded (or so he thought), it would surely get him the recognition and rewards that he craved.

The problem was that this concept is just woefully wrong for the wiring harness business. There are some very low-volume car assembly plants that kit materials for vehicle assembly, and the kitting of materials for aviation harnesses for final assembly made sense in Alabama due to the extremely low volumes and the processes involved. But, the kitting of materials for final assembly for automotive wiring, even for low-volume production, makes zero sense.

When this director rolled out his kitting ideas to me (I had engineering responsibility for the plants) and the materials manager, we tried very hard to convince him that his concept was a bad idea; that it would add a tremendous amount of labor cost in the warehouse to prepare the kits, would reduce the productivity of the plants due to extra servicing and the significantly increased potential for missing materials, and would harm quality as the extra handling of the wires would potentially cause some damage, and it would also be much easier to misidentify components.

Unfortunately, none of these very valid arguments swayed him at all. He was convinced (or so he said) that this was a leaner system, because there would be less materials in the plant (which is true, but there was not less material in the system), and it would provide for better material flow (which is nonsense, in addition to the fact that sending kits when assembled is not even a true pull system).

The existing method for supplying the Mexican plants with materials, since the implementation of the nonsensical remote cutting and lead prep system (already described in Chapter 3), was for each plant to place an order to the Mexican Distribution Center (MDC) whenever an item (a lead code, a plastic component, a metal component, etc.) reached its reorder point. The MDC would then pull the required amount of material and send it to the plant by truck (each plant received at least one daily shipment). I always felt the distribution center was a very costly, nonvalue-added facility, but with the logistical nightmare of getting materials from several different Warren supplier plants to several different Mexican user plants, there was probably no better option. It was just another of the very big costs associated with remote cutting and lead prep. Regardless of the cost, this system worked reasonably well, with the caveat that remote cutting and lead prep was a disaster overall.

I shutter even today when I think about the system that this director put into place; the nightmare we had to live with while trying to start up two new plants, which is a big enough challenge in itself. First, he established a kit size of 200 pieces, not because any line had an actual requirement of 200 pieces, but because this was about an average per shift output on an average line. However, that's like saying the average weight of a Sumo wrestler and a ballerina is 300 pounds. Some lines produced 100 harnesses per shift and some produced 600 or more harnesses per shift. There were also several common lines in the plant, and we had several situations where six kits were sent to supply four conveyor lines, or three kits were sent to supply two lines.

This created more than enough logistics problems to last a lifetime, but the biggest problems we had were in how the kits were received. In the traditional system, each tub of material received would be one specific material code.

There were no multiple codes in a tub, and the material had not been touched since it was processed in Warren and placed in a tub. With the new kitting system, everything changed. It was impractical to designate a tub for only 200 pieces of wire or for a component, so dozens of different material codes were placed in the same tub.

Believe it or not, this is how the system was designed to work: in the distribution center, a "mother" tub of each of the needed materials was lowered from the racks with a fork lift (unless that material happened to be on a lower level); 200 pieces of material were removed and placed into a "kit" tub (normally either 2 or 4 bundles of wire) on top of the material already in the tub, until the tub was full and another tub was started; and then each of the mother tubs was returned to the racks. For components, mother tubs were lowered, 200 pieces were weigh counted and placed into plastic bags that were then put into "kit" tubs, and then the mother tubs would be returned to the storage racks.

You are probably thinking that this is a heck of a lot of nonvalue-added labor even if you are doing several kits for the same conveyor line(s) at the same time, and you would be right. But, it was even worse, because the decision was made to complete one kit in entirety before starting on the second, supposedly in an effort to maintain control. In any case, I'm sure you get the gist. The bottom line is that scores of workers were added to the distribution center in order to assemble the kits. And, then the kits were sent to the plants, with significantly increased shipping costs due to more frequent deliveries.

When the kits arrived at the plants, the real fun started. Instead of having one location within a final assembly department to receive and store a specific material (which was in pristine condition because it had not been handled since it was processed), it was now necessary to dig through kits to find an urgently needed item. Even in the best case, when there were no material shortages, which allowed each material to be taken out of the tub in the reverse order it was put into the tub, it was a challenge getting the material to the line that had the greatest need, because there was an imbalance between what was in the kits and what each line required. It was also a very frequent occurrence that one or more items were missing from a kit. In fact, it was an exception to receive a kit with everything in it. This led to many cases where leads had to be moved back and forth between lines to balance out the inventories in an effort to keep the lines running until the next kit arrived. Unfortunately, the next kit often did not arrive in time.

You also can imagine the condition of the materials after all of the additional handling to which it was subjected. Leads which were to be later put into splices had one or more stripped ends, and the stripped ends were almost universally frayed after all of the extra handling. This resulted in serious quality problems, as well as labor penalties, in the splice-making areas. Some terminals also are fairly delicate, and the additional handling did create some plugging problems that generated extra scrap.

In order to cope with all of the extra handling requirements foisted on the plants due to the kitting fiasco, several additional workers had to be added within the plants as well, and still it was almost impossible to deal with all of the logistical problems created within each plant. The lines were up and down like yo-yos, and there was great fear that we would not be able to meet customer delivery requirements. In an effort to try to get our materials under control, a decision was made within our plants (without the knowledge or approval of this director) to install conventional racks in the back of the department, move the kits to the back of the department when they arrived, and then put the materials into or onto the specifically designated storage location. Because kitting was much more of a push system than a pull system, eventually there was a buildup of inventories within the plant, and, with the newly installed racks, it was possible to keep materials controlled well enough to keep the lines running and keep GM supplied. But it was nip and tuck for several weeks.

To summarize, just so you get a clear picture, Packard Mexico added scores of nonvalue-added workers to the Mexican Distribution Center to make kits that were sent to the various plants, and then, within the plants, this material was taken from the kits and placed in individual storage locations (with dozens of additional nonvalue-added workers), all the while doing significant damage to the quality of the materials. Believe it or not, this system remained in place in some of the Mexican plants for a period of several years. While the organization finally learned how to make and control kits well enough that we were not constantly endangering the customer, it certainly endangered the bottom line.

When you consider the tremendous labor increase in the distribution center, the additional material handling costs in the plants, the additional transportation costs, and the reduced conveyor productivities due to line shut-downs due to missing materials, millions of Packard dollars were flushed down the toilet. And why? Because a director wanted to make a name for himself by implementing a system that somebody else, somewhere else, in some other type of business, said was a good idea (see sidebar).

By the way, I think the best system we could have used to supply the plants from the distribution center, and the one I kept pushing (to the point of jeopardizing my career), is the very simple and easy to control two-tub system that functions with the use of a two-tub flow-through rack system. When the first tub is emptied, the second tub slides into position and the empty tub is the signal to pull more material from the warehouse. Inventories can be easily controlled for each type of material by the quantity of material put into each tub. Why this extremely simple and effective system was never put into place within the Mexican Operations is a mystery to me; we certainly pushed for it. While my only direct experience in Material Control occurred when I had total operations responsibility in both South Korea and Alabama, this seemed like a no-brainer to me; at the very least, very worthy of a pilot trial, but no dice. It would have saved millions of dollars in just the first two plants in which we were trying to implement kits, not to mention all of the others that were forced to follow.

Let's Do the Team Concept Like Toyota

This same director also had one other bright benchmarking idea, but I do know where he got this one. He had traveled to the NUMMI plant and had observed the team concept, among other things. Of course, the folks at NUMMI raved about the system, and this director's reaction was: If it's good enough for NUMMI, it's got to be good enough for us as well. After all, if you've seen one business, you've seen them all, right?

Well, the simple fact is that all businesses are not the same, and what works well in one may or may not work in another. It is certainly worth taking a good look at an interesting concept, doing a thorough evaluation, and initiating a pilot if it looks promising. However, don't be too disappointed if something that works great in another company is not suited at all to yours.

Because of Toyota's adherence to their first sacred rule (All work shall be highly specified as to content, sequence, timing, and outcome), they have a big competitive advantage over their U.S. rivals. In essence, Toyota preplans its assembly lines such that each operator has a well-defined station having a work content time that is very near the established line speed (i.e., each station has very little line balance or waste). This is not the case with most of its competitors. Therefore, Toyota is able to assign a team leader for every seven or eight operators, who do such things as cover for absenteeism or emergency breaks, provide training, help resolve quality problems, sort material, and a host of other things, and still be in very good shape versus its competition.

Fortunately, at Packard, through the flexibility provided by off-line operations (see Chapter 5), we could accomplish exactly the same things without going to the expense of adding team leaders who would have limited productivity during much of the day. When he explained his concept of team leaders, I reminded him of our flexibility and ability to accomplish what Toyota accomplished with their team leaders, but at a much lower cost than by adding unnecessary heads. Again, my arguments did not fit his agenda (of implementing new and exciting and innovative concepts), so "Team Leaders" it was.

The Team Leader Concept turned out about how we knew it would. The team leaders contributed virtually no productive hours to the plant, and the things that they did accomplish could have been done very effectively at a fraction of the cost through utilization of tried and true systems. After a few months, the manufacturing management of the plant scrapped the whole team leader concept (again without telling or getting the approval of the director), thereby saving tens of thousands of hours of labor per year. They only talked about and gave the appearance of utilizing the Team Leader Concept when they knew the director was coming for a visit. It was a shame that they had to use this deception, but what choice did they have?

This is not to say that the team leader concept in not valid and will not save money in certain businesses and in certain situations, but, it is certainly not a panacea. This concept, like virtually all others, needs to be looked at with intense scrutiny and evaluated on the benefits that can be derived versus the costs and the risks associated with it. There have been many apparent successes with the team leader concept, but there seem to have been just as many dismal failures, based on my readings and investigation. Don't dismiss the concept out of hand, but don't think that just because Toyota does it, it is bound to be the right thing for you to do as well.

How about a Nice Quality Circle?

I'm not sure how quality circles came into such prominence. Toyota has used them on occasion to deal with specific problems, but even W. Edwards Deming, who was greatly revered within Japan for the work he did primarily on quality, was not a big fan. He said about quality circles, "That's all window dressing. That's not functional. That's not fundamental. ... Anything goes wrong, do something about it; overreacting, acting without knowledge, the effect is to make things worse."

I'll be the first to admit that I am not unbiased about quality circles. I have had very good success solving quality problems, not to mention other types of problems, in the plant by utilizing the experience and knowledge of hourly workers, who sometimes have the most insight into problems dealing with the operations to which they are assigned eight hours a day. But, these have not been done in formal quality circles, which I have never observed to have functioned very effectively. It has been done through interviews, observation, questioning, and discussion. Rarely, if ever, was any productive time lost through my efforts, and I'm confident the operators involved did have a positive feeling about being asked for their input and participating in problem solving.

My only personal, first-hand experience with quality circles was in South Korea. Our Quality manager had received prior training in Total Quality Management, and he was a convert to the concept of quality circles. He twisted my arm for several weeks, and I finally agreed to let him set up a few quality circles to deal with some issues in the plant, with the agreement that we would get report-outs in a couple of weeks. At the end of two weeks, a formal meeting was set up and each of the teams (five or six) made formal presentations of their experiences and what was accomplished.

The bottom line was that hundreds of labor hours were lost, expectations were raised, and absolutely nothing was accomplished that would make a difference in the plant. In fact, resources were taken away from legitimate efforts to resolve the problems (with input from the hourly workers) and put into trying to facilitate the quality circles. So, my personal experiences were pretty negative to begin with. They only became more so after I observed some of the things going on in the Mexican plants.

I'll only give one example, of the many that I could give, that demonstrates that Packard looked at these Lean tools as "ends in themselves," and not as a "means to an end," which is the case with Toyota. After returning to the States from South Korea, I toured several of the plants to refamiliarize myself with the Mexican Operations and some of the new things being done. In one of the plants, I was taken to an information center of which they were intensely proud. It was very elaborate, with lots of charts and graphs and pictures, and it would certainly be impressive to most visitors. I scanned several of the charts and something jumped out at me very quickly. It seemed that something had happened about a year and a half earlier that had a very negative impact on their plant productivity. Their productivity was expressed as a plant-wide efficiency percentage based on estimated labor content for each product, so it was not all that helpful to identify and resolve specific problems, but the labor estimates did tend to be relatively constant for similar packages, so comparing one year's productivity to another's should certainly be order of magnitude.

I didn't ask about the big drop in efficiency at that time, I wanted to see more of the plant first. But, I assumed that I was going to find out that they had changed products within the plant, had undergone a gigantic model change, or something else with equal or greater impact had occurred. While on the plant tour, we passed several conference rooms within the plant. I was told that these were the conference rooms used by the Quality Circle Teams. Apparently, they had undertaken a very big effort to incorporate the concept of quality circles into their plant operating philosophy. Each department had at least one quality circle, and several departments had two or more circles. The idea was to rotate departmental employees through the various quality circles, so that everyone had an opportunity to participate. When one circle would wrap up its work, another would be formed, so that there were ongoing quality circles in each department at all times.

Obviously, I was curious as to when these quality circles had started and what evidence they had that they were paying dividends. Low and behold, the quality circles had started at precisely the same time that the productivity charts started showing a big drop in plant efficiency. Upon further questioning, I found out that the business within in the plant had not changed for several years and the plant had not experienced a big model change for at least three years. When asked about the benefits that had been derived from the quality circles, no one was able to give any concrete examples, just a comment that the operators seemed to really enjoy the experience. When we were leaving the plant, we passed by the information center again. This time I looked at the quality charts, and, as you've probably guessed, they didn't show any improvement over the same time period either.

It seems that the only noticeable impact the plant received from its obsession with utilizing quality circles as an end rather than a means to an end, was a dramatic decrease in productivity and a corresponding increase in plant costs. Again, this is not to denigrate quality circles and not to say that they never have a place, but, it is to say that it is a tool, not an objective, and when this tool is used, it should be targeted for a specific purpose, with specially selected individuals, with specific objectives in mind, and in a controlled environment and time frame.

Did QS9000 Implementation Really Help?

In the mid-1990s, the Big 3 U.S. automakers decided that one of the things they were going to do to become more competitive was to implement QS9000 (which was a list of quality- and manufacturing-related procedure requirements), and to insist that each of their suppliers become QS9000 certified or risk losing business. I do not plan on getting into a discussion as to whether or not QS9000 (which may soon change to ISO/TS 16949) was a good thing, or whether or not it was worth the effort. Looking at the current sad state of affairs of the U.S. automakers, it would make one wonder, but it might be a case of "in spite of" and not "because of." What I will tell you is that the implementation of QS9000 certainly was not worth the effort within Packard, based on the way it was implemented, with the exception that it was necessary in order to maintain business.

If nothing else, QS9000 gave Packard the opportunity to standardize systems and procedures within the various plants, which was sorely needed at that time. Every plant had its own way of doing things. It was almost a point of pride for a plant that management did things differently and looked different from every other plant. I voiced my opinion repeatedly that this was a very unhealthy and costly situation, one which made it difficult to share resources, human or otherwise, between plants. What I would say to anyone willing to listen was that there may be a lot of ways to skin a cat, but, if we wanted to be effective and save money as an organization, we needed to settle on one of these ways and implement it universally. If someone could demonstrate later on that they had an improvement that saved real money, improved quality or delivery, or generated some other measurable benefit, it should be adopted as the new standard, which every plant should then be required to implement. Anything done outside of the standard, with the exception of a carefully controlled pilot, should be looked at as a negative and corrected back to the standard, and not held up as an example of innovation and creativity.

However, this was not done. You would think that a company that appeared at times to be obsessed with Lean Manufacturing would do everything possible to standardize. This is certainly a Lean concept, but this was not to be. What actually occurred is that all 26 Mexican plants worked independently on their QS9000 certifications; this was actually little more than documenting what they were currently doing, making modifications or additions as necessary to meet the requirements, and then submitting these documents in the formats required.

The benefit that Packard got out of this very arduous, time-consuming, and expensive process was exactly zero. If it had been used as an opportunity to determine the best practice (or even the selection of one of the various good practices) for each of the elements for QS9000 certification, and then document and implement each of these practices throughout the entire Mexican Operations, some good would have come from the effort. As it was, other than maintaining current business, no good came of it, quite the contrary. It was a colossal waste of money. No wonder the U.S. auto industry is in the shape it is in.

Say Zero Defects and Mean It

The concept of zero defects also became part of the Lean dialogue within Packard. It became heresy to suggest that "zero defects" was not a reasonable goal, or that it was not actually achievable. It's kind of like saying that hitting a home run every time at bat was a reasonable expectation, and that to say anything to the contrary was to risk being branded a heretic.

But, the truth of the matter is that the great quality guru, W. Edwards Deming, who was credited with much of the quality success that Japan enjoyed during the past few decades of the last century, preached to Toyota and others that there are acceptable defects and an acceptable defect rate. Rather than waste efforts on zero defect goals, Dr. Deming stressed the importance of establishing a level of variation, or anomalies, acceptable to the customer in the next phase of a process. Often, some defects are quite acceptable, and efforts to remove all defects would be an excessive waste of time and money.

The distinct impression I got from virtually all within the Packard's Quality Organization was that there is no such thing as an acceptable defect rate or spending too much money to eliminate a defect, even though, as Deming stated, there are defects that do not concern the customer, or that he or she will even notice. For example, there are many dimensions on a wiring harness that are not at all critical, and through design reviews at the assembly plant and through prior experience, it is not difficult to ascertain which ones are critical and which ones are not.

Many of the Packard plants, in an effort to "improve" quality, would actually use an audit board to inspect a large percentage of the harnesses, and management would instruct the operators to "repair" harnesses that had noncritical dimensions out of tolerance. If the dimension was short, the "repair" would usually involve the operator pulling and tugging, or, if the dimension was long, it would involve the operator tucking and taping. In any case, the cure was worse than the bite. It not only cost money to inspect for and "repair" these noncritical dimensions, but what would otherwise be an unnoticeable dimension variation to the customer might become a noticeable problem created in the "repair" process.

However, the biggest cost-add related to the concept of zero defects occurred because of a very questionable decision made on how the electrical test would be performed. Sometime in the 1990s, a decision was made that affected most of the plants to the effect that a single ring-out of all of the circuits in the harness was no longer sufficient to ensure quality.

The argument was made by the Quality Organization that it was conceivable for an electrically defective harness to pass the electrical test if it was only rung-out one time. However, this could occur only if that harness was repaired on the ring-out board and only if the supposedly highly trained ring-out operator was to (for some unexplained reason) remove a circuit that had already been tested as good, replug it into the wrong position, and then ring-out the rest of the harness.

To the best of my knowledge, no one had ever known of this happening (and for it to happen, it would almost have to be intentional), but Quality insisted that the only way to ensure that it would never happen would be to ring-out each harness completely two separate times. If a rering only took a few seconds, this might have not been too big a problem; but on a big harness it might take several minutes. The end result was that we had to add scores of additional operators to perform the double ring-outs, as well as spend hundreds of thousands of dollars in extra floor space and electrical test equipment, all in an effort to fix something that was not broken. I'm sure someone in the Quality Organization thought they had come up with a real innovative solution to an imagined problem.

I often told my engineers to give themselves the following challenge when evaluating whether or not something should be done: Pretend that you will reap all of the rewards of the new concept, but also pretend that it is your money that will be used to fund it. Using these criteria, Packard would have avoided most of the big mistakes we have already reviewed, most especially this one and a whole lot more.

The Good News Is That We Saved $25 Million in Inventory, but ...

I have already mentioned that most of the times when Packard was encouraged by (i.e., ordered by) GM to cut costs, the cutting was virtually always done in two areas: the first being headcount reductions and the second being inventory reductions. It is not that either of these things should never be reduced, it is just that reducing either of them to solve a short-term problem in a forced exercise rarely works out well, especially when the cuts are made in a vacuum without consideration of the potential impact on other interrelated factors within the company.

Every manufacturing system is comprised of many elements, each having its own cost and performance factors, and each being related to several of the other elements within the system. A list of these elements would include such things as: direct labor, indirect labor, salary labor, overtime, quality, delivery, efficiency, floor space, utilities, maintenance, tooling and equipment, raw material inventory, in-process inventory, finished goods inventory, transportation, scrap, rework, inspection, travel, taxes, supplies, and benefits. When a decision is made that is directed toward one of the elements in the system, it will virtually always have an impact on other elements in the system. Because of some of the problems I had observed over the years at Packard, I took a word that had been used in another context and used it to describe an activity in which Packard had engaged several times, with very negative consequences.

I used the word "suboptimization" to refer to the activity of focusing on a single element in a manufacturing system with the intent of obtaining optimum or maximum performance for this single element without considering the impact on the entire system. When this is done, it is very possible to create a situation in which the performance of the element on which focus is placed, in fact, does improve, but because of the spillover influence on other interrelated elements within the entire system, the performance of the entire system suffers a deprovement (a word used in Industrial Engineering (IE) to mean "it got worse"). This does not mean that because this potential does exist, we should use it as an excuse to not vigorously pursue improvements through waste elimination. Where true waste exists within a system, it must be identified and eliminated. In most cases, if it is true waste, it can probably be eliminated without having a negative impact on the entire system, although it may require that some aspects of the system be redesigned which are not in harmony with truly Lean concepts.

One of the worst examples of suboptimization happened in Mexico in the late 1990s and involved the suboptimization of inventory. Packard management determined that the best way to save money, yet again, was to dramatically reduce inventory. Because prior efforts had been somewhat in vain, the determination was made to establish a firm target against which Mexico would be measured and the implication was that there would be no excuses for not meeting the target, which was established as a $25 million inventory reduction. I don't remember exactly what percentage this amounted to, but I am thinking it was in the area of 20%.

The problem was that no systemic changes had been made that would provide any assistance in reaching this target. If the target was going to be met, it would be through grit and determination, as opposed to establishing new systems and procedures to make it more feasible, such as eliminating remote cutting and lead prep, which would have made it a snap, but that just wasn't in the cards.

At the end of one year, a Lean tour was conducted in one of the Mexico West plants, during which several Lean projects would be reviewed, including the status of the Inventory Reduction Project (IRP). After the wide-aisle tour was complete, the dog-and-pony show began. One of the first presenters was a good friend of mine, who was given the opportunity to showcase the IRP. The problem was that he is only a little less politically correct than I am, so he decided to tell the whole truth and nothing but the truth.

His presentation went something like this: "I've got some good news and some bad news about the IRP. The good news is that we did hit our target of reducing inventory by $25 million. Assuming that our carrying cost for inventory is about 10%" (*the carrying cost takes into account the opportunity cost of money and the cost to store and manage inventory; it would be much lower than 10% today*), "this means that we were able to add about $2.5 million to the bottom line. Not bad. However, the bad news is that our premium freight bill for Mexico increased by $10 million this past year as a direct result of having insufficient materials in the plants to keep the manufacturing systems running and our customers supplied with products in a timely manner."

You could have heard a pin drop. The Mexican director finally said something like: "We'll have to take a look at it," and we moved on to the next act in the show. At the end of the show, I went up to my friend and congratulated him for speaking the truth, although I was sure it wouldn't do his career any good. But, I also informed him that the situation was much worse than what he had presented. He had only mentioned the $7.5 million net loss due to premium freight. He did not mention the additional $10 million loss incurred due to a 5% decrease in productivity throughout the Mexico West Operations due to conveyor lines being down because of missing materials.

The bottom line is that we spent (wasted) $20 million (a very conservative estimate) in order to optimistically save $2.5 million in reduced inventory costs, and, in the process, we also created a much more negative working environment, and made it much more difficult for the supervisors to run efficient departments, even when sufficient materials were available. When lines are up and down like yo-yos, workers get the distinct impression that busting their tails to keep the lines running just isn't that big a priority. Their thought process was just like yours or mine would be in this situation: "The line is going to be down before long anyway, so why kill myself?"

This is not to say that inventories should not be reduced. They should if more material is being inventoried than is necessary in order to optimize the entire manufacturing system. But, remember that even Toyota, recognized for inventory control, uses inventory as a countermeasure to deal with the realities of a manufacturing system.

Just-in-Time Manufacturing?

As the old saying goes, there is a time and a place for everything, and this is certainly true in the case of JIT manufacturing. Packard considered this concept to be the epitome of Lean, even though when Toyota talks of JIT, it refers primarily to inventory control and material flow and not to manufacturing, as such.

However, JIT manufacturing does have its usefulness. In fact, going back to the early 1970s, Packard utilized the concept of JIT manufacturing frequently in its manufacturing processes; many executives just didn't know it because JIT was not yet on anyone's radar. The ultimate example of JIT manufacturing was the integration of the splice presses into the conveyor lines (which, of course, was foolishly terminated as part of Vista when JIT was a buzz word within Packard). Even when splices were made off-line, they functioned very much like a JIT manufacturing process in that completed splices were delivered to the line at frequent intervals. Subassemblies also were frequently made at or near their respective conveyor lines and funneled into the lines as they were made or at frequent intervals. There were also many other examples of JIT manufacturing processes used to supply other manufacturing processes.

Apparently, these examples of JIT manufacturing processes were just not sufficient to satisfy Packard executives in the late 1990s. This could have been because a lot of the JIT manufacturing processes we had used so successfully in the past had been discarded such that little JIT manufacturing was left. Unfortunately for Packard, in 2001, shortly after I had left the company, a consultant was able to convince Packard to implement JIT manufacturing in the one area in which it should never even be considered in a wiring harness plant, and that was on the final assembly lines.

How a consultant, who had no real experience in automotive wiring, could convince Packard to do such a foolish thing is beyond understanding. Even if it could be feasibly done, there was absolutely no advantage in doing so. Based on customer orders, it was only necessary to ship each part number once or twice per week. But, the big problem was that it could not feasibly be done. Good manufacturing supervisors knew that the way to improve quality and productivity within the department was to minimize the number of part number changeovers on each conveyor line during the week, hopefully to the point that a given part number would not be run more than one cycle during the week. Every part number change required a change of identification tape, a change in the ring-out sequence on the line or on the off-line ring-out boards, a change in the methods of some or all of the operators, a change in the packaging container, and, frequently, even a change in the line speed.

The one factor that was more responsible for quality and productivity problems than any other in a final assembly department was part number changeover on the line, regardless of what systems or processes were put into place to minimize the impact. Therefore, the right answer was to minimize part number changeovers, not maximize them. But, under the direction of a consultant, maximize them they did, and it occurred in one of the plants for which I had at one time provided support.

What the consultant convinced them to do was to build the wiring harnesses in the order that the assembly plant intended to use them sometime in the future. I can't even imagine the impossible task of building each of the harnesses correctly and getting them into the proper container. The sad thing is that the plant manager actually happened to be one of the better plant managers in Mexico West at the time. I just don't think he had that much say in whether or not this new concept was going to be implemented in his plant.

I found out about this tragedy when I happened to bump into this plant manager sometime later when he was working for a different company. It seems that he was the fall guy when this new concept went south, as anyone who had any experience in wiring harness manufacturing knew it would. Because of the unbelievable problems that were created with trying to implement this new system at model change, the unpardonable sin of actually shutting down a car assembly plant occurred. Someone had to fall on his sword, or have it done for him, so the plant manager did, or had it done to him. While this plant manager should have never allowed this to happen, I suspect that the real blame for forcing this debacle on his plant lay elsewhere.

Yes, car assembly plants do build each car to a specific set of specifications, but wiring harnesses are not cars. Don't make the very invalid assumption that a concept that might work well in one company making one type of product is necessarily transferable to another company making an entirely different product. Just ask Packard what the consequences can be if you do.

I'll finish this section with a quote from one of the more enlightened consultants used by Packard over the years: "There is not a single answer for Lean Manufacturing. Many consultants have fostered digital thinking for Lean manufacturing: single piece flow, zero inventories, coupled processes, no process islands, no buffers, etc. There are logical exceptions to all of these Lean Manufacturing rules. There is no substitute for thinking." To this, I would add: **Do what is smart, not what a consultant or a book tells you to do.**

All I can say is "amen," although I would submit that, in my experience, it would be more the exception than the rule that many of these so-called Lean concepts are advantageous.

U-Cells Gain Great Prominence

Over the years, Packard had done a great job of utilizing conveyors and a good job of utilizing batch processes, where they made sense, within the various manufacturing systems. As the years rolled on, after the introduction of Lean and TPS in the mid-1980s, batch processing became more and more stigmatized (even though it was far and away the most cost-effective solution for much of the processing that needed to be done within the various plants). The mere mention of the word "batch" brought stares of disdain from many Packard executives.

In order to be process pure and avoid any appearance of clinging to the archaic concept of batch processing, the new concept of U-cells was introduced in the late 1990s in several of the plants by one of the up-and-coming Mexican superstars, with encouragement from another consultant. Whereas before, various subassemblies had been produced in an area containing, as an example, two terminal presses, one sleeve applicator, and four splice presses, Packard would now install four U-cells to produce these assemblies. This meant that we would now have to buy two more terminal presses and three more sleeve applicators, which we would use in conjunction with the four splice presses, to create the U-cells (meaning that the three pieces of equipment would be laid out in a U-shape configuration).

A piece of material would start at the terminal press and, after being processed, it would "flow beautifully" to the sleeve applicator, where it would be processed by the same or a different operator. Then it would flow beautifully to the splice press station, where it would be processed by the same or a different operator, who would then dispose the subassembly into a container (see sidebar).

In reality, the first utilization of U-cell technology occurred in the assembly of BECs or Bussed Electrical Components, and there were actually half a dozen or more types of equipment involved, but the concept was exactly as described here.

I heard about this concept in an executive meeting, and I expressed some concern that we would not be able to recoup the cost of the extra equipment and floor space. It seemed as well that we might be adding extra labor and material handling to the process. I was assured that this was a Lean system that would reduce inventory and improve material flow and, because of this, it would be more productive. I just grinned and thought to myself: "Here we go again."

It took about six months (which was longer than I expected) when a modification to this system was announced in a similar executive meeting. With the help of the consultant, they had come up with the name "Multicell" for the new modified system. They had found that it was just too costly (who could have seen that coming?) to have a dedicated piece of each type of equipment in each of the U-cells, because the utilization was very low for several of them, not to mention the loss in labor productivity due to so much extra material handling. So, what they decided to do was to determine how many units of each type of equipment were needed to meet requirements, then they would set up a large U-cell containing the required equipment. The material would flow from equipment type #1, to equipment type #2, to equipment type #3, and so forth to the last type of equipment, where the subassembly would be complete and packed in a container.

At the end of the presentation, I asked something like the following: "Excuse me, but I'm a little confused. Can you please explain how what you are describing differs from the concept of batch processing that we have employed for decades?" After a lot of stammering, the answer was something to the effect that this was a different concept and the material flow would be different. I didn't say it, but I was thinking, "Yeah, it's just like the difference between a janitor and a sanitation engineer, or it's the same thing as calling someone short or vertically challenged." It reminds me of another saying we used a lot at Packard: "If it looks like a duck, walks like a duck, and quacks like a duck, it probably is a duck." This was definitely a duck, and a very expensive duck at that.

Somebody Finally Got One-Piece Flow Right ... Really?

I could give a lot more examples of very "Lean" attempts by Packard (meaning, attempts without substance or reason) to implement Lean concepts and to become a Lean organization, but I am sure you are getting the picture. However, I wanted to finish with the coup de grâce. This situation occurred in a plant where we produced harnesses for a non-GM truck customer. Because of the nature of the truck business, our estimating system was not very accurate, so the plant ran at relatively low efficiency, but this was taken into account in the pricing formulation so that this plant was actually one of the more profitable plants in Mexico West. It should be noted that the cost per standard hour that Packard used in pricing non-GM business was much lower than what was used for GM business because the impact of Warren was not considered in the cost profile of non-GM business. In other words, GM was subsidizing the harnesses we were selling to non-GM customers. I'm sure they would have been very happy to have known this.

Before I get into the example, I want to give a little background that will be helpful in understanding how such a thing could happen in a mature manufacturing organization. If Toyota is noted for anything, it is noted for its focus on the elimination of waste. Through their intensive efforts to understand and to find waste, Toyota was able to identify seven forms of waste that are potentially present in all manufacturing operations. Their belief is that if these types of waste can be identified and then eliminated within a manufacturing process, tremendous improvements are possible.

Toyota's seven forms of waste include:

1. Inventory: Any supply in excess of process requirements necessary to produce goods and services just-in-time. (*And I would add, any supply in excess of the optimum lot size as determined through scientific analysis.*)
2. Overproduction: Producing more than is needed or sooner than is needed.
3. Correction: Nonvalue-added inspection and repair of a product or service to fulfill customer requirements.
4. Material movement: Any movement of material that does not directly support a synchronous system. (*Or I would add, any movement of material that does not support optimum lot size flow.*)

5. Processing: Effort that adds no further customer value, enhancements that are transparent to the customer, or work that could be combined with another process.
6. Waiting: Idle time that is produced when two dependent variables are not fully synchronized.
7. Motion: Any movement of people or machines that does not contribute added value.

In total, I agree with Toyota's list, as long as these three caveats are included for Packard:

1. Very frequently within Packard (in fact, almost without exception), controlled batch processing was by far the most effective means to provide processed materials from cutting and lead prep, as well as from subassembly and component operations (a controlled batch process being one in which lot sizes have been determined through scientific analysis to be the most effective for the system as a whole).
2. These forms of waste are not mutually exclusive; in fact, many are interrelated in most manufacturing processes. For example, in an effort to minimize inventory, we may significantly increase waiting time or processing time or motion.
3. There is one very significant category of waste which I believe must be added to the list of seven, which is the following: **Underutilization of resources, underutilizing tooling, equipment, floor space, labor, etc. due to poor planning or imbalances in the manufacturing system.** If this very tangible form of waste had been considered by Packard, many of the poor decisions would not have been made (hopefully).

Another thing that entered into the decision to proceed with the example I will be giving is Toyota's notion of the ideal manufacturing system, which many at Packard bought into hook, line, and sinker. As described by Steven Spear and Kent Bowen, for Toyota, the output of an ideal person, group of people, or machine:

1. Can be produced in a work environment that is safe physically, emotionally, and professionally for every employee
2. Can be produced without wasting any materials, labor, energy, or other resources
3. Can be supplied defect free

4. Can be supplied on demand in the version requested
5. Can be delivered immediately
6. Can be delivered one request at a time (which meant, according to Packard, a batch size of one)

I'm glad Toyota refers to this as the ideal system and not the realistic system. Within a well-managed Packard plant, the first three conditions on the list are certainly approachable, even if they will never be completely accomplished. The fourth and fifth can be accomplished with reasonable inventory levels, but this creates some conflict with the second condition. But, the real challenge is with condition number six.

Perhaps in another universe, if the material handling time is zero, travel time is zero, the setup time and changeover time is zero, and there are no quality or productivity concerns related to extremely frequent product changeovers, a batch size of one might be the ideal. But, here on Earth, this is absurd for virtually every process within any Packard facility. The problem is that because Toyota said it (regardless of the context in which it was said), it must be true.

This being the case, Packard developed a principle around this concept and made it a feature in the Packard Production System. They wrote the principle as follows:

Single piece continuous flow is the least costly method of providing value.

Believe it or not, this was widely believed within Packard, even though we had countless cutting and lead prep processes (supporting conveyor lines) that had output requirements of several hundred to several thousand pieces of identical product on a daily basis. I knew there would be a lot of confusion over this principle, so I rewrote it and presented it to upper management to try to get their buy-in. I rewrote the principle as:

Single piece continuous flow IN THE FINAL ASSEMBLY PROCESS is (generally) the least costly method of providing value, i.e., a conveyor, a U-cell, or an assembly line.

Unfortunately, I could not convince Packard to change the way the principle was worded. It had something to do with not dampening the enthusiasm of employees implementing the "Lean" concept of one-piece flow, even if it made virtually no sense anywhere except on our conveyor lines, which were already a world class example of *proper* one-piece flow, even though Packard executives didn't seem to realize this.

With this background, we can proceed to the example I referred to at the beginning of this section. The plant manager involved in this fiasco was actually an experienced engineer (although I do not believe he had any manufacturing experience, other than as a plant manager). However, he also was intently interested in furthering his career. He had already come up with some gimmicks in his plant that looked kind of neat, although I doubt they saved a nickel. But he knew that if he could come up with a system that demonstrated that he truly understood the principle of one-piece flow and a batch size of one, the sky was the limit.

The system he devised was truly amazing, and I don't use that term lightly.

It was obvious that our current cutting machines and lead prep processing machines were totally incompatible with the concept of a batch size of one. The only feasible way to manage a cutting or lead prep department was to set up and run reasonable batch sizes, certainly no smaller (and probably a lot larger) than one day's worth, because even raw materials were delivered on a once-a-day basis. Since our excellent conveyor system was already an example of a continuous single-piece flow system, he knew that somehow he would have to configure cutting and lead prep differently if he was going to achieve any notoriety for being the guru of one-piece continuous flow.

What he did for each conveyor line was to design a series of U-cells that jutted out from the side of the line. Each U-cell was designed to produce the leads that were required for two or three stations, so that on an Instrument Panel line, there might be four or five U-cells supplying leads. Because it wasn't very practical to try to put four or five large cutting machines adjacent to a line, another system of cutting leads had to be utilized. What he came up with was a manual cutting and stripping head mounted to the top of a small table with a sliding scale to be used for controlling cutting lengths. So, instead of using automatic cutting and terminating machines that could process a day's worth of requirements in a few minutes, he was proposing going to a manual cut and strip (but not terminate) of each wire on a table, with a batch size of one (kind of how we made one-piece samples in the Methods Lab).

No, the operator was not going to cut 50 or 100 leads to make a bundle before passing the bundle on to the next station. That just would not do for a single-piece, continuous flow process. The operator was to cut a single lead, walk over to the solder pot that comprised the next station, solder that single lead, then dispose it to a tray, and return to the cutting table to repeat the process.

Probably few of you that are reading this are wiring harness manufacturing experts, but I think you can understand from what I have shared that this concept not only made no sense at all, it bordered on insanity.

The next process was terminating. Since each of the U-cells would be supplying several different leads with an assortment of different terminals, it was necessary to find a way to minimize setup time. A turntable device was devised that would hold at least four terminal application dies, and the appropriate die could be rotated into position rather quickly when a die change needed to be made. This was kind of a neat concept, except that it took up a lot of room and was quite expensive, and it would never be utilized in a conventional lead prep area because there was just no way to justify it. Usually, a second operator was used to do the terminating. This operator would take a lead from the tray and apply a terminal, then pass the lead on to the splice machine station or directly to the conveyor line, whichever was pertinent.

I first saw this system when I went over to the plant to see the new system I had heard about through the grapevine. What I had heard made no sense at all, so I had to go check it out. I had a hard time believing what I saw. It not only bordered on insanity, it was insane. It was obvious that what this plant manager was proposing would at least quadruple the labor cost of each processed lead, would almost double the space utilized by the conveyor system, and would add hundreds of thousands of dollars of additional investment in tooling and equipment, with the only tangible offset being a few square feet of storage space in the back of the final assembly department. Oh yeah, I almost forgot, we also could lay claim to a process that was a true-to-life continuous one-piece flow (that is, if you didn't want to count all of the paced operations—the epitome of one-piece flow—which we already had throughout the operations that actually made sense). In essence, what was being proposed was like going from Star Wars back to the Stone Age.

After seeing what was being done, I diplomatically told the plant manager that I thought it might be tough to justify the system. He was so blinded by the desire to implement the process that he just could not look at it objectively. I suggested to him that if he insisted on pursuing the project, at the very least he might want to consider cutting a bundle of leads before passing them on to the next station instead of sending a single lead at a time. Although he appreciated my input, he countered that doing this would not represent a true one-piece flow, which was of paramount importance.
I left him with my best wishes, especially because a big tour of his plant by Packard's director of Lean (Packard really had such a position) and several other Lean honchos was scheduled in a couple of weeks.

The day arrived and I felt sorry for this plant manager. He was a nice guy who was only trying to do what he thought his superiors wanted, and what he thought would get him ahead. I figure that he was going to get his head handed to him—or so I thought. I guess I should have known better by now.

I made the decision that I was going to try to do the impossible of keeping my mouth shut during the tour so that I did not unduly influence anyone else. I didn't think it would be necessary because what was being shown was obviously a disaster in the making. I started to get a sick feeling when I began hearing a lot of ooohs and aaaahs when the system was explained to the group. Surely they couldn't think that this concept made any sense, could they?

After the plant tour, there was a wrap-up meeting with the plant manager, his entire team, the entire Lean entourage from Warren, and most of the mid- and upper-level managers from Mexico West. What I heard in the meeting was unbelievable. The director of Lean praised the plant for being the first to really understand the concept of one-piece flow and to do something about it. He stated that this plant would be named a Five Star Plant and would be used as a showcase for others within Packard to come and learn about Lean Concepts, especially concerning one-piece flow. They also informed the plant manager that he would get the approval for the $7 million he needed to convert the entire plant to this concept. I just sat there thinking, "Are you kidding me? No one could be this crazy."

That evening I went in to see my boss (another suit), who was the director of Engineering for Mexico West, and asked him what he thought about the tour. He indicated he was quite impressed with the tour and anxious to see the plant converted to the new system. I asked him how he thought we were going to be able to pay off the $7 million investment, because we would be quadrupling cutting and lead prep labor, not to mention doubling floor space requirements, and there were no offsetting savings that I could see. He just said he was confident that the one-pieced flow system would increase productivity sufficiently to pay off the investment in less than the projected two years. Could he really be this blind, or did he just have to support the team plan? I told him that I couldn't disagree more, and that if we proceeded with the plan, we would regret it—and soon.

I decided then and there, because I already had 30 years of service, that if I got a chance to retire early I would. A company that would make this kind of decision didn't have a very bright future. That opportunity came about six months later when a special early retirement was offered, and I took it.

Over the preceding six months, I had been keeping up with the performance of the plant as the new system was being slowly implemented. Each month since the Lean Tour, the plant performance and the plant profitability had been steadily digressing, as more and more of the plant was converted to the new system. I refrained from saying "I told you so" and decided to wait and see when and what steps would be taken to stop the bleeding.

When I told my boss that I was taking the special early retirement, he was not altogether shocked. He knew how disillusioned I had become. To his credit, he did say, without any prompting on my part, that I was probably right about the one-piece flow system, which was a tough call because the plant productivity had decreased by 50% (everything was out of control since so much effort was needed to try to make the new system work) and the plant profitability had decreased from 15% to a negative 5%. I told him that I appreciated him telling me so and said nothing else. What was the point?

I left Packard soon thereafter, so I am not sure what finally became of the new system, which was only about 50% implemented in the plant at that time. I have to believe they stopped the implementation and reverted back to conventional technology, but with the kinds of decisions being made at Packard, who knows?

Again, the question has to be asked: "How could intelligent people, with a lot of experience in wiring harness manufacturing, be so wrong and make such a terrible decision?" It should have been obvious to anyone, regardless of experience, but especially to this group, that the system being proposed was just plain NUTS, with a capital N. There were no savings to be realized, only additional cost, and, it required a $7 million investment to achieve these additional costs.

I have to believe that there were others who were skeptical, although no one ever spoke up or let me know that they were in agreement with my assessment, but they understood that there was nothing they could do, so why jeopardize their careers. Once the Lean director said that this plant was the first Packard plant to truly "get" the concept of one-piece flow—a pillar of the Packard Production System—the die was cast.

Wrap-Up

I hope this chapter has been a real eye opener for those of you who are currently engaged in (or plan to get much more involved in) Lean Manufacturing techniques. I could give a lot more examples of Packard's futility in trying to implement Lean concepts, because there just was not a clear understanding of what was supposed to be accomplished, but I think what I have shared should give plenty of warning that great care and good judgment need to be exercised. When Lean concepts are implemented as an end in themselves rather than as a means to an end, you have a recipe for disaster.

Eliminating waste, which is the primary goal of Lean, is a goal that any manufacturing company, in fact, any company, should pursue vigorously. But, it must be remembered that the seven forms of waste, as defined by Toyota, in addition to the eighth that I added, are not mutually exclusive. What is critical is to reduce the overall waste of an entire system, not just the waste of one or more of the elements within a system, without considering the impact on the system as a whole.

For the last example that I gave, it could be argued that three of the forms of waste (correction, over production, and inventory) were reduced and possibly optimized. However, the other five forms of waste (motion, material movement, processing, waiting, and underutilization of resources) were all dramatically increased. The overall impact, without even full plant implementation, was a dramatic reduction in plant performance in virtually every category. And, why was this horrific project approved and implemented? Because one of the pillars of the Packard Production System said: "Single-piece flow is the least costly method of providing value." How naïve and how costly can one simple statement be?

Chapter 5

Proper Manufacturing Organization Is Critical (*or* You Will Get the Performance You Motivate)

When I first started working for Packard in January of 1971, it had been in operation as a manufacturing company for longer than virtually any of its employees had been alive. During this time, a lot of smart people had worked for the company and, undoubtedly, a lot of different organizational structures and concepts had been tried in both the manufacturing and in the support organizations in an effort to improve the productivity and cost effectiveness of the operations. One would think that in a period of seven or eight decades, smart people would gravitate to the right answers on at least some aspects of the business (through a process of trial and error) such that no more improvements could be made, assuming the business incurred no significant changes.

I observed that this is exactly what had transpired in establishing the functional support departments within Packard and establishing their relationship to the manufacturing departments. In Engineering, for example, each of the Engineering disciplines (i.e., Industrial Engineering, Process Engineering, Product Engineering, and Plant Engineering) had a responsible superintendent who reported to an Engineering director. This provided continuity and consistency within each of the Engineering departments, as well as providing for ease of training and communications, among other things.

However, it was clearly understood by each of us in nonmanufacturing functions that we were there to support manufacturing, which was drummed into us as being the only functional area in the entire company that produced a product that was sold to a customer and generated income. All of the rest of us were taught that we were a burden, and, if we did not provide services to the plant that provided more benefit than what we were costing, our existence could not be justified. Because of this understanding, when I was not engaged in preplanning to establish a manufacturing system for a new product, I spent virtually all of my time in the plant (along with all the other Industrial Engineers [IEs]) helping to identify and solve production and quality problems, thereby helping to drive down costs and justify our existence.

In trying to recall the organization charts from those days, in addition to seeing a solid reporting line from me to my functional supervisor. I don't remember seeing a dotted line to manufacturing supervision, but, this is the perception I had, even if it didn't show up on the organization chart. Each of us IEs understood that we were part of the plant operations' team, comprised of manufacturing representatives and representatives from each of the functional areas. We knew that without our full support, manufacturing could not be successful, and we knew that feedback from manufacturing, based on our level of support, would have a big impact on our yearly appraisals and ultimate success with the company.

Plant Priority Meetings

One of the tools that was used very effectively at Packard at that time, which further emphasized the operations team, was the Plant Priority Meeting. These meetings were conducted weekly under the chairmanship of the plant manager or his designate, and there were key representatives (supervisors or their designates) from each of the primary support organizations, i.e., IE (the IE supervisor coordinated and conducted the meeting and maintained the control sheet), Process Engineering, Quality Control, Material Control, Scheduling, Maintenance, and other areas as needed to speak to particular issues or problems.

Each issue or problem for which the plant needed functional support to resolve or accomplish was brought up and discussed in this meeting. After sufficient discussion, a priority was assigned to each problem/issue and one of the team members was assigned responsibility to use whatever company resources necessary to get resolution. Each issue continued to be reviewed in succeeding meetings until a satisfactory solution or resolution was in place, then the issue was removed from the list.

These meetings were conducted on a year-round basis, and they were very successful in providing a format for getting things done within the plants. Every functional area was made aware of plant needs. These needs were properly communicated and coordinated, and the issues and problems were resolved in a timely manner.

When we started up the Mexican Operations, there was no plan in place to utilize the successful concept of Plant Priority Meetings. I recognized that this would invariably lead to reduced plant performance, because the support areas would not be as engaged as they should be and would not be providing the most effective support possible. I took my concerns to the Manufacturing manager, who was a very smart guy, but who had not had any manufacturing supervisory experience at either the first or second level. He seemed to be somewhat sympathetic to my concerns, but he made no efforts to implement these meetings, even when I offered my full support in helping to get them started.

The bottom line is that these meetings never became part of the culture of the Mexican Operations, much to our detriment. It was a simple tool that would have paid big dividends. When I wrote the IE Manual, I included a section on manufacturing support and one of the things I included was the need to conduct regularly scheduled Plant Priority Meetings. I have attached the format, with instructions, from the manual (Figure 5.1). I would encourage any manufacturing company to utilize this type of tool to greatly enhance performance. In fact, this type of tool should be used to provide coordination, communication, and control of important issues in virtually any type of company, manufacturing or otherwise.

Let's Just Change Reporting Lines to Get Better Support

As I suspected, things did not run as smoothly in the first Mexican plant as they could have due to less than ideal communications between the plant and the support areas (although, for an initial startup in a new country, things went better than most expected). These communication issues could have been solved with something as simple as regularly scheduled Plant Priority Meetings, but the Manufacturing manager had another idea. His idea was a concept that I had heard mentioned a time or two earlier (in the late 1970s).

PLANT PRIORITY MEETING From Instruction

LE SUPERVISOR	①	MFG. MANAGER	MAINTENANCE	MANUFACTURING	MATERIALS
PLANT MANAGER		INDUSTRIAL ENG.	MES	QUALITY CONTROL	SCHEDULING
	②	DAY _____ TIME _____ A.M./P.M. LOCATION _____			

Priority	Description of Area/Activity Needing Improvement	Responsible	I.P.?	Assigned Commit Actual	Status	Remarks/Follow-up
③	④	⑤	⑥	⑦	⑧	⑨

◐ Define & Assign   ◐ Develop Plan   ◐ Begin Implementation   ◐ Implementation Complete   ● Follow-up Complete

1. The Multi-Discipline Plant Priority Team is comprised of at least one responsible individual from each of the operational areas that are supporting and running the plant. These representatives will normally be supervisor level, if available.
2. The meeting is scheduled weekly at a specific time and location. All areas must be represented and the representatives must come to the meeting prepared to discuss the issues to which they are assigned.
3. Each issue put on this form is to be given a priority, i.e., high, medium, or low. The issues should be sorted in this priority. Within the priority, sorting should be done by commitment date, with the nearest commitment date coming first on the list.
4. Describe the problem or improvement need in some detail. This list should include issues *(from all areas)* that are having a negative impact on plant performance.
5. A specific individual *(or at the least a specific area)* is indicated, who is most responsible to resolve the issue.
6. Indicate whether or not a specific Improvement Plan *(formalized multi-discipline project)* is in place or is scheduled in order to solve the problem, such as a Six Sigma Project, a Lean Workshop, etc.
7. Indicate the date that the issue is officially assigned to the responsible individual or area, the date the responsible individual or area commits to have the issue resolved, and the date the issue is actually resolved. Once the issue is resolved, and follow-up indicates that the resolution is irreversible, the issue is removed from the active sheet the following week. Resolved issues should be placed on a separate format that will contain at least a one-year history.
8. This column is used to show a visual representation of the status of each of the issues. Updates are done weekly.
9. Make any other remarks necessary to further clarify the issue, what needs to be done, the status of resolution activity, and/or the follow-up status.

Figure 5.1 Plant Priority Meeting control sheet.

Apparently it was an idea proposed by some guru, whose credentials probably didn't include real-life experience on a manufacturing floor.

His idea was that the best way to improve in-plant coordination and cooperation, so that things got done in a timely and effective manner, was to have some of the key support people report directly to manufacturing, with dotted line reporting responsibilities to their functional heads. Apparently, to some, this sounded like a great way to create an effective operations team that would work together to get things done. Manufacturing would no longer have to entice support from other functional areas. The individuals providing the needed support would be on the team with direct reporting responsibilities.

However, I could see that this type of organizational structure was destined for big time trouble down the road. So that there could be no misunderstanding about what my concerns were, I wrote a letter to the Manufacturing manager, to the Engineering manager (my boss), and to the other Engineering supervisor who would be affected. The letter was dated October 27, 1978, which was a few months after we had started production in the first Mexican plant. Below is the letter, which I am including because it still contains valid information for anyone who might be considering such an organizational change in an effort to improve communication, coordination, and performance (but will do just the opposite).

> I am writing this letter in response to the proposed engineering reorganization for Mexican plants. I believe the advantages this organizational change is supposed to create can be accomplished by other means without the potential disadvantages inherent in the proposed system.
>
> Your proposal, as I understand it, is to have the Industrial Engineers and the Process Engineers, who are assigned to various manufacturing modules within the plants, report directly to the Manufacturing module managers, while having dotted line reporting responsibility back to their functional supervisors in Central Engineering. I believe the things you are trying to accomplish with this reorganization are:
>
> 1. Provide the module managers with total operating control within their respective modules.
> 2. Create an engineering and manufacturing team within each module.

3. Provide the framework for the team to work jointly on module objectives with unity of purpose.
4. Create the understanding that performance will be based on accomplishment of these objectives.

I completely concur with all of these desired results; however, I submit that they can be accomplished without a drastic reorganization of engineering. My suggestion is that a minimum of one Process Engineer and one Industrial Engineer be assigned to each module, based on size, but that these engineers continue to report to their respective Process Engineering or Industrial Engineering supervisors within the plant. These engineers would be part of a module team having responsibility for establishing and accomplishing module objectives, and their performance would be based in large part on the accomplishment of these objectives.

My proposal is very similar to systems currently in operation in several Packard plants, and it is the general consensus of the general foremen in these areas that this system works exceptionally well. Engineering responsiveness to problems is excellent and engineers are considered key members of the team that has responsibility to set and achieve objectives.

I believe there are several inherent problems associated with the proposed organizational change—problems that would be very difficult and costly to overcome, if they could be overcome at all. A partial list follows:

1. This would create too great an area of responsibility for one man. Running an efficient production area is a difficult enough assignment without having the additional workload of trying to supervise two dissimilar engineering disciplines.
2. Finding qualified module managers would become a very difficult assignment; very few individuals are competent in as many as two of these areas—much less all three. The module manager, most likely, would be incapable of providing adequate training, supervision, and insight into resolving engineering problems, or of even realizing and recognizing shortcomings on the part of his or her engineers.
3. No checks and balances would exist within the plant. IE has always been the guardian of costs, but can that be the case if

they report to a manufacturing supervisor (can you say "conflict of interests")? Process Engineering also needs to be in a position of being able to make tooling and equipment decisions without undue influence.
4. No means would exist to provide consistency within the plants. My fear is that each area would make engineering decisions within their respective areas, independent of the other areas. Industrial Engineering and Process Engineering methods, standards, systems, and, most especially, tooling and equipment, would become bastardized in a short period of time, leading to untold problems in future years.
5. A very unhealthy competitive atmosphere could develop. A fear is that each module, in an attempt to outdo the others, would keep improvements within their own module and not share beneficial information and concepts with others, or, even if they wanted to share improvements, there would not be an effective mechanism to do so.
6. No coordination or communication vehicle would exist for IEs or for PEs for information coming from the central engineering areas in Warren and Mississippi. Each plant will have the need to coordinate such things as: model change schedules, board tooling, engineering changes, pilot programs, tooling and process changes, advanced or new engineering techniques, etc.
7. There would be no effective coordination or control of the engineering training programs within the plant.
8. True engineering skills could be lost. A fear is that engineers could become little more than manufacturing "gofers" if short-sighted module managers do not have the proper perspective and long-term vision in how to best solve problems for the long-term success of the operation.

My humble opinion is that potential problems 1, 3, 4, 5, and 8 would be very difficult, if not impossible, to solve, and, if we proceed with this new organization, we will face serious long-term problems in these areas. Potential problems 2, 6, and 7 could possibly be solved within a central engineering group, but I believe this would be a very difficult and awkward task and that it would be far less effective than having IE and PE supervisors within the plant, to whom the plant assigned IEs and PEs would report, with dotted line reporting responsibility to the module managers.

I have written this letter because I am very concerned about this issue. I believe very strongly in what I have written and what we have discussed. I have no personal consideration in what I have suggested—to the contrary, I would have more to gain personally through an increase in the size and scope of Central Engineering (which this new proposal would dictate), but I firmly believe that the proposed reorganization is ill advised.

After receiving my letter, the Manufacturing manager (who happened also to be a good friend of mine) called me into his office. He told me that he appreciated my concern, but he said he believed that I had overstated the potential risks and he was confident his proposed system would work well (after all, he had read it in a book). I reminded him again of the benefits of implementing Plant Priority Meetings as a means to improve communication, coordination, and performance within a plant, but he insisted that his new plan was the way to go.

I talked to my boss, the Engineering manager for the Mexican Operations, and his position was that we were there to support manufacturing, and, since this was what they wanted Engineering to do, this was what we were going to do.

I probably don't need to tell you how things worked out; not very well would be putting it mildly. It didn't take all that many months before the ill-advised change was reversed. So, from the standpoint of long-term damage, this fiasco was only a blip on the radar screen. What was not a blip on the radar screen was a fatal mistake that was made concerning how manufacturing departments would be organized and managed.

I have written about a number of very expensive mistakes that Packard made over a relatively short period of time. The mistake I am going to share with you now (concerning how manufacturing departments were to be organized and managed) was the most devastating of all; a mistake that surely haunts Packard to this day. This mistake, along with mistakes I will detail in the next chapter, was responsible for Packard wasting multimillions of dollars annually by failing to properly control its most important and costly asset—its workforce (in this case, specifically direct operators).

There are certainly no guarantees that Delphi could have avoided bankruptcy if they had avoided the mistakes from this chapter and the next, seeing how most of their economic woes are due to stateside union problems, but, if these problems had been avoided by Packard, Delphi would have been hundreds of millions of dollars better off.

Warren Got Some Things Right

As is evidenced from this book, Packard made a multitude of very big mistakes over the years, even in the early days, but, over a period of many decades, you would expect that they would get a few things right. Well, they did and they will be discussed in the rest of this chapter and in the next two chapters, namely, manufacturing department organization, efficiency control (direct labor control), and preplanning.

When I say that Packard got some things right, I am not implying that they used these tools, processes, or systems as effectively as they should have, or that they did a good job transferring these concepts to new operations around the world, they didn't. I am not even suggesting that they realized the value of these jewels that had been developed. They obviously didn't or they would not have allowed these concepts to disintegrate. What I am saying is that Packard developed three systems (inclusive of the respective tools and processes) that were world class in the early 1970s, concepts that would stand the test of time (allowing for updates using modern technology), which are just as applicable today and will be tomorrow as they were at that time. The fact that Packard allowed two of these three systems to crumble and the third to badly deteriorate is a travesty.

Understanding Human Nature

Based on the way departments were organized within Packard in the early 1970s, as well as evidenced through the productivity control system, it was clear that some of the early Packard managers understood human nature a lot better than most of the managers that came along later. I'm not sure why, except that as the world became more touchy-feely as time went by, objectivity about human nature seemed to diminish. Early Packard managers seemed to understand the concept that "people will do what they are motivated to do," or, said a different way: "You will get the behavior you motivate." These concepts have always been valid and always will be, whether we are talking about behavior within the family, any organization, the state, the nation, or the world, and they are especially valid if anyone is trying to determine how to set up and run a manufacturing department.

In the rest of this chapter, I am going to be referencing wiring harness manufacturing departments, but the same concepts are applicable to any type of manufacturing department (or nonmanufacturing department, for that matter).

A typical Final Assembly Department at Packard in the early 1970s usually contained somewhere between 30 and 60 direct operators who were under the supervision of one foreman. This typical department would contain a couple of conveyors, each utilizing 10 to 20 operators, and several off-line operations. The off-line operations were frequently stationary boards, which built the very low volume part numbers of the conveyor package (so that the lines would not have to be changed over so often), subassemblies that supported the line, splice stations that built splices which supported the lines or stationary boards, a repair station, and/or other stationary boards or build stations.

There were several different classifications within each Final Assembly Department, thanks in large part to a short-sighted union and weak-kneed management. Two classifications would have been ideal, three would have been good, but the five we had were at least two too many, especially with the restrictions that existed on how hourly workers could be used. However, even with having to deal with the inflexibility of the operator classifications, most Final Assembly Departments were well organized and productive (i.e., they utilized direct labor efficiently, produced good quality, and met delivery requirements).

1. "Operator" was the first classification. These were the lowest paid operators in the department, who were each paid the standard union rate and made up about 60% to 70% of the department employees. (It was actually a good thing to have this many employees with a single classification in order to facilitate control within the department; promotions could be used to recognize superior operators.) These operators were assigned to the various stations on the conveyor lines.

    The rest of the classifications earned a small premium above the standard operators, only a few cents per hour; however, there was a lot of competition for these other classifications (the ones not tied to seniority), primarily because of the increased flexibility and responsibility that these jobs offered. However, not all operators were interested in these more responsible jobs (a surprise, no doubt, to a lot of experts).

Many employees were more than happy to come into work, work almost robotically in their station all day, and then go home at the end of the day without worry and with a fat paycheck.

But the operators who had an interest in doing something more challenging and varied competed strongly for these positions.

2. The "Universal Operator" (they were actually called Extra-Girls, but I will use the term Universal Operator because it is more descriptive and there were also men in some of these slots starting in the early 1970s) was the most important hourly classification in the department, and this group of operators was largely responsible for whether a department ran smoothly or not. When there was no absenteeism in the department, they would be assigned to the off-line operations. They usually worked it out among themselves to rotate among the assignments so that everyone's skills remained sharp and no one got bored. The expectation—and the reality—was that each Universal Operator would meet the established standard on the job to which he or she was assigned each and every day (or else they knew they would not be able to keep the classification and they would be back on the line).

When there was absenteeism on the lines, it would be covered by a Universal Operator, each of whom knew every station on both lines, and it was done in a democratic fashion. They maintained an equalization list that showed the number of hours each of them had worked on the line since the list's inception (new Universal Operators were assigned the average number of hours for the group). The low person on the list automatically knew he or she would be the first to go onto the line, so, at a couple of minutes before the bell, he/she would determine who was missing and he/she would make his/her way to that station. If two line operators were absent, the second lowest Universal Operator on the list would take that station, etc.

A good foreman wouldn't even have to show up for work and still be confident that his/her lines would be started on time and that the Universal Operators would look at the department schedule, see what jobs needed to be worked, and then start doing them; but good supervisors were always there, anyway. Universal Operators also were used for special assignments, such as special cleanups, sorting of defective materials, and training, and they were used as well to cover absenteeism for any of the other classifications in the department.

This was a critically important group of operators in any Final Assembly Department. Good Universal Operators meant a smooth running department, poor Universal Operators meant a department with a lot of problems and an overworked and harried foreman. Fortunately, foremen had the right to promote operators into this classification (and remove them from it, if deemed necessary). This ensured that the best of the best occupied these critical positions. That is why I was so stunned when (sometime in the 1990s) Packard management gave away this right in contract negotiations. One thing the Union cannot tolerate is performance-based rewards for the workers covered under its contracts. This was just another nail in Packard's coffin; a very big and serrated nail, at that.

3. The "Relief Operator" was the operator who filled in when someone needed emergency relief. Because line operators were tied to the line, except when at lunch or on breaks, if a line operator had a legitimate emergency, someone needed to step in for them so that the line would not be stopped (a big no-no). Unfortunately, as always happens, line operators started abusing this system. Weak foremen had their Relief Operators tied up eight hours a day; strong foremen did not allow emergency relief for the first hour after startup, breaks, and lunch. That way we at least got four hours of productive time from our Relief Operators.

4. "Service Operators" (they were actually called Service Boys back then) were responsible for putting away all materials coming from the raw material store and from the cutting and the lead prep departments. They then serviced this material to the lines or off-line operations as needed to keep all processes running. There was normally enough space in the work stations to accommodate between two and four hours of material, so each lead and component needed to be serviced to the proper location two or more times per shift.

5. The "Repair Operator" was responsible for repairing any defects that could not be repaired easily on the line, which was always the first option when a defect was detected. A good supervisor would drive the number of defects which came off of the line to practically zero (this was done over time, as each defect would be systematically analyzed for root cause and corrective action). This meant that a lot of Repair Operators were underutilized, unless they could be loaned out to other departments.

The ideal situation would have been to have only two operator classifications in final assembly: Operator and Universal Operator. The Universal

Operator would then have been assigned to all of the functions, except for those listed for Operator, which would have given the foremen much greater flexibility and would have led to improved department productivity. However, even with this handicap, in those days, good foremen ran very well organized and efficient departments.

Each Packard department was set up to have enough Universal Operators to cover maximum absenteeism (excluding some catastrophic event) and to perform off-line functions that needed to be done daily. Engineering was responsible to ensure that each department had sufficient off-line operations to ensure a productive workstation for each Universal Operator on days when there was no absenteeism within the department, while also ensuring sufficient capacity within these operations to allow for normal absenteeism coverage.

Normally, things worked out well. Adjusting inventory levels was normally the first tool used if there were temporary labor imbalances, the loaning or borrowing of operators from another department was normally the second tool used, the use of overtime was the third, and, occasionally, short work weeks or reductions in headcount (over and above attrition) were necessary. But, by and large, Final Assembly Departments (as well as Lead Prep Departments, which were organized somewhat differently due to the nature of the work) ran exceptionally well, especially when having to deal with a not so cooperative union.

So We Transferred Our Manufacturing Successes to Mexico, Right?

One would think that if something was successfully being done in one manufacturing location within Packard, strong consideration would be given to introducing it into new operations as well. Unfortunately, this was not the case regarding the organization of manufacturing departments. These important concepts were not introduced to Mexico (they had not even been transferred to Mississippi, as I later found out).

Why weren't they introduced? For one thing, there was no one named to the 10-man startup team who had any manufacturing supervisory experience at either the first or second level other than me, and I was busily engaged in a lot of other activities. When I did step out of my area of responsibility and offered advice to the manufacturing manager, my advice was usually not warmly received.

In my mind, the most critical position on the startup team was the Manufacturing manager, someone who should have had a deep and thorough understanding of manufacturing with a proven track record. That team member just did not exist. The person responsible for manufacturing received his very limited experience at the superintendent level, and that is just too far removed from the floor to really understand how departments should be organized and what works and what doesn't. It wasn't his fault that he did not have this experience and understanding, although, he should have been astute enough to realize that he needed expert assistance in this area and asked for it, but this didn't happen.

There were a couple of prevailing attitudes within Packard at that time among top executives, which I think were largely responsible for not getting a key manufacturing expert on the team. The first attitude was that we were going to Mexico because things were so messed up in Warren and they wanted a start fresh in a new environment. (However, not everything was messed up; we actually knew what we were doing in the manufacturing organization.) The second attitude was that since labor was so cheap, some of the manufacturing controls that were in place in Warren were no longer necessary. (They couldn't have been more wrong. As I explained, even with the low wage rates, by far the most expensive component of cost in any plant is direct labor.)

So, what happened with the manufacturing organization in the first plant startup? First of all, the departments were laid out almost identically to what would have been done in a typical Warren Final Assembly Department. This would include the requisite number of off-line operations to provide workstations for all of the Universal Operators within the department on days with zero absenteeism. Without the union dictating company policy, we only established two classifications within the Final Assembly Departments. This should have provided the Mexican Operations with tremendous flexibility to run effective manufacturing departments. Unfortunately, that's about where the similarities ended.

Mexican foremen, who were largely left on their own to figure things out, deduced that since most of the product they were required to build in order to meet shipping requirements came off of the conveyors, they should put their best operators on the conveyor lines. Conversely, because a relatively small amount of production came from off-line operations, operators who could not (or did not) perform well on the conveyor lines were assigned to the off-line operations. The thinking was that these poor-performing operators would create a lot less damage in the off-line operations than on the conveyor line on which, if they couldn't keep up, they would have affected the output of the entire line and everyone on it. With this plan, they also realized that they needed to hire additional operators to cover absenteeism, because the operators working on the off-line operations were not qualified on any of the conveyor stations.

Some of you reading this might be thinking: "Well, this sounds pretty reasonable to me. They had to get product shipped out the door, didn't they?" Clearly, but before we give them the benefit of the doubt, let's consider the characteristics of both conveyor (paced) jobs and off-line (nonpaced) jobs in Table 5.1.

Now, put yourself in the shoes of a Mexican operator who has worked exceedingly hard and effectively on the conveyor and expects to be recognized and rewarded for those efforts. What do you think his/her attitude will be when the person who is working next to them on the line (who misses work regularly, can't keep up on the line, and makes frequent defects) is taken off of the line and assigned to one of the off-line stations? What would your attitude be and how would you respond?

If you answered that this operator would likely be very demoralized and his/her performance would probably deteriorate, you are correct; and this is exactly what happened throughout the Mexican operations in each of the new plants as they came on stream, and it continues to this day.

There are two things that are as certain as day follows night: (1) if you want more of something, you reward it, and (2) if you want less of

Table 5.1 Characteristics of Paced vs. Nonpaced Jobs

Characteristics of Paced Jobs	Characteristics of Nonpaced Jobs
Easy to learn and to become proficient	More difficult to learn and gain proficiency
Easy to control	More difficult to control (for both productivity and quality)
Successive verification easy	Successive verification difficult
Physically demanding	Less physically demanding
Minimum variation	Significantly more variation
Tied to the line	Flexibility of movement
Perceived low status	Perceived higher status
In summary, not very desirable	Much more desirable

something, you penalize it. For example, if you want operators to perform better, you reward good behavior and you punish bad behavior. If you do the reverse, i.e., if you punish good behavior and you reward bad behavior, the results are totally predictable and disastrous.

In essence, without really understanding it, the Mexican supervisors were rewarding the bad behavior of the operators who could not (or would not) get the job done on the line by assigning them to relatively cushy off-line jobs where their performance was not really monitored and they were not hassled. At the same time, they were punishing the good operators, who were doing everything asked of them, by keeping them on the line in much less desirable jobs than their peers who were doing a miserable job.

What was the end result? Do you really have to ask? Of course, the outcome was quite predictable. The result was that the Mexican Operations Final Assembly Departments had productivity levels that were about half of those in Warren, and this was in spite of, and not because of, the influence of the Warren union. Without the negative influence of the Warren union and the restrictive classifications of the labor agreement, the gap would have been larger. About 10% of that gap was due to high turnover rates in Mexico and another 10% due to material shortages, especially after the introduction of remote cutting and lead prep. The remaining 80% was due to the factors explained above plus the impact of eliminating the productivity control system, as explained in the next chapter.

As you might imagine, there became fewer and fewer standout operators; there was nothing in it for them. Most of the operators who did a great job, at least when they started and before they became jaded, ended up on the line and there they stayed. The only way they would be rewarded for their excellent performance was to be assigned as a Universal operator responsible to cover absenteeism, and there weren't that many opportunities. Even when they got one of these opportunities, it meant that they would have no productive job to do when absenteeism was light, and, when there was high absenteeism, they would be back on the line filling in.

Operators who really disliked working on the conveyor quickly learned that their best avenue for escape was to perform poorly. This could well earn them a ticket to a plush off-line job. Operators who were assigned to the off-line operations performed even worse than they had on the line, where there had at least been some peer pressure and a little supervisory attention because the lines had to run reasonably well to meet shipping schedules.

Everything was 180 degrees out of phase, and this was not an isolated situation; it was universal within the Mexican Operations. It may well be that the right way to organize a department just isn't intuitive. However, why let all of the knowledge Packard had gained in over seven decades of refinement go to waste?

I did a study to try to gain more insight into the great disparity between the productivity in Warren and that in the typical Mexican Final Assembly Department. In Warren, I knew that most of the conveyor lines made the standard on a daily basis (ran at 100% efficiency), virtually all Universal operators met the standard every day on the off-line operation to which they were assigned, and there were no people floating around the departments without productive jobs. By way of contrast, Mexican operators assigned to off-line operations were performing at efficiency levels well under 50%, most conveyor lines were producing in the neighborhood of 75% or 80% efficiency, and there were all kinds of people milling around the departments with no productive jobs to keep them busy. The answer was pretty simple, really.

What's the Fix?

So, what needed to be done to fix the mess? Mexico productivity levels in final assembly were half of those in Warren, including the fact that Mexican lines had to be programmed with at least 20% to 30% higher capacity than Warren lines in order to meet delivery requirements. It's the same answer that almost any coach gives to a reporter when asked why his or her team just got blown out by another team: "We've got to get back to the basics."

Just what were the basics? The basics were the tried and true concepts on organizing and managing a Manufacturing Department, which had been refined in Warren over many decades, and which had proved very successful, even when having to deal with a less than cooperative union. Mexico had the advantage of not having to deal with a union, so it was not necessary to create additional classifications that hindered flexibility and productivity. Mexico did not have a union that would go to the mat for employees who deserved to be fired. In Mexico, if an employee deserved to be fired for very poor performance (especially within the first 90 days), it could be done quickly and with little difficulty. In Warren, if someone deserved to be fired, it was extremely time consuming and laborious to make a reality, so, generally, foremen just learned to deal with the individuals as best they could.

In an effort to help fix the problem, I strongly suggested that the following things be done in Mexico immediately:

- All new hires should be automatically assigned to a conveyor station. If any of them were not able to perform on a conveyor station (which was known to be a fair station) in a reasonable period of time after having received proper training, they should be dismissed and allowed to find employment elsewhere.
- All operators assigned to off-line stations should be transferred back onto the line in an orderly manner. If they continued to be unable or unwilling to perform at an acceptable level, they also should be given an opportunity to find employment elsewhere.
- The very best operators should be identified and trained as potential Universal operators (to include the functions of relief, repair, and service, which were done by separate classifications in Warren), such that they are able to run each of the conveyor stations and make the standard on each of the off-line operations (just like they did in Warren). Those who made the grade should be promoted to Universal operator and taught the remaining aspects of the job, such as how to use the equalization sheet to cover absenteeism, how to effectively rotate among the off-line jobs, etc.

There was a desire on the part of several key individuals in Manufacturing and in Human Resources to create three classifications of operators in Manufacturing. This also could easily be incorporated, and, in some regards, created an even more effective organization within the Final Assembly Department, as well as in the Cutting Departments and Lead Prep Departments.

For example, Level 1 operators, comprising somewhere around 55% of the operators in Final Assembly Departments, would be assigned to the conveyor stations (excluding the last station, which was the most critical station, because this operator was responsible for inspecting each harness as it was being removed from the line, and then he/she folded each harness and packed it into the proper container).

Level 2 operators, comprising around 30% of the operators in the department and who would be paid more money than Level 1 operators, would be assigned to the last conveyor station, to the electrical test station (if the electrical test was not done on the line), to the service function, and to off-line operations that needed to be run on a daily basis. These operators could rotate jobs to keep skills current and to make their jobs more interesting.

Level 3 operators, comprising around 15% of the operators in the department and who would be paid more money than Level 2 operators, would be assigned to cover any absenteeism within the department, train new operators, accomplish the relief and repair functions, and work on the off-line operations not required to be run every day (most especially stationary boards). The bottom line is that they would have a productive job to do every day, with the exception of the time they spent training other operators.

All of the jobs for hourly operators in Lead Prep and in Cutting were more desirable, but more difficult to control, than conveyor jobs, so most of the jobs in these departments should be Level 2. Level 3 jobs would be reserved for operators capable of meeting the standard on every process within the department, and, therefore, they would be capable of covering for absenteeism or providing training as needed.

When an opening came available for a Level 2 operator (in Final Assembly, Lead Prep, or Cutting), the most qualified Level 1 operator (as determined by the foremen involved) should be promoted into that position and a new operator hired to replace him/her. When an opening came available for a Level 3 operator, the most qualified Level 2 operator (as determined by the foremen and maybe the general foremen/supervisors) should be promoted into that position, which would probably leave an opening for another Level 2 position that should be filled by the most deserving Level 1 operator, etc.

Let's reflect back on the concept that "you will get the behavior (or performance) that you motivate."

- What is the likely impact on an operator's performance when he or she understands that doing a lousy job is going to earn him or her a one-way ticket out the door instead of a ticket to a better job?
- What is the likely impact on an operator's performance if he or she realizes that if he or she does a really good job in all aspects, he or she can earn a promotion to a more enjoyable job with more pay? (In Warren, the pay differential was minimal between an operator and a Universal operator, but the competition for these jobs was fierce, none the less.)

I won't insult your intelligence by answering these questions. It's the same answer you would receive anywhere else in the world, in any kind of business or endeavor. Why were these concepts so difficult to grasp for many of Packard's managers? It might have something to do with not thinking straight when alligators are nipping at your heels, especially when your level of knowledge and expertise are not what they need to be, and you are unwilling to listen to others who might be able to help.

Manufacturing Supervision Is Tough

When I was in the Philippines, working primarily with Industrial Engineering and the Methods Lab, I did have occasion to spend quite a bit of time in the various plants. In Chapter 10, I will review some of my experiences and recommendations regarding their Cutting Departments, but I also observed some significant failings in their Lead Prep and Final Assembly Departments as well. It was obvious that they needed a lot of help understanding the basics of how to properly organize and run a manufacturing department. I put some thoughts together for them (see below). I had a chance to provide some instruction before I left (**most of which are universal truths**), mostly to the plant managers, and the feedback I received is that things started to improve dramatically as some of these concepts were put into place. Some of these ideas and concepts might be of benefit to you readers as well even though they were specifically written for wiring harness manufacturing in the Philippines, although almost everything is applicable to wiring harness manufacturing anywhere.

Providing Proper Manufacturing Support

Industrial Engineering, more than any other support organization, is tasked with the responsibility of helping manufacturing reduce the plant cost per hour by reducing expenditures on the Plant Operating Report. While the Plant Operating Report contains many items of expense (maintenance, supplies, allocations, scrap, etc.), about 70% of a typical plant's expenditures are in the labor accounts and generally about 70% of a plant's labor account is for Direct Labor (the remaining 30% being for Indirect and Salary Labor).

This means that approximately 50% of a plant's nonmaterial-related expenses are for Direct Labor. (This surprises a lot of people who have always thought of Mexican Direct Labor as being cheap.) However, when you consider that almost every other line item on an Operating Report is directly or indirectly influenced by Direct Headcount, **the actual impact of Direct Labor on our nonmaterial operating cost is probably 70% or higher.** Because Industrial Engineering is responsible for establishing good methods, layouts, tools, and material flows (along with providing associated standards and proper training) through preplanning and other analyses and evaluations, it is clear that the proficiency with which we do our jobs will have a big impact on the cost of Direct Labor in the plant.

However, there are other ways that a Methods and/or Industrial Engineer can provide support to Manufacturing. Manufacturing can be a difficult and challenging assignment. To be effective in Mexico, a Manufacturing supervisor must manage a large group of high turnover employees with limited education; achieve acceptable levels of productivity, quality, delivery, and responsiveness; and keep expenses to a minimum so that he/she can meet the budget and attain profit targets.

Because the Methods and/or Industrial Engineer should be working very closely with the Manufacturing supervisor, he/she can help ease the burden and provide proper support if he/she understands a few very important concepts about the manufacturing organization.

While the quality of the job done by the Industrial Engineer has a significant impact on the Direct Labor cost (as mentioned before), the biggest impact on Direct Labor cost (aside from what we pay our Direct operators) is the productivity level of our Direct Labor. It can be demonstrated that the manner in which departments are organized and the ways operators are assigned and controlled are the biggest influences on productivity.

With this understanding, it is important that Methods and Industrial Engineers know the important concepts on how departments should be organized so that they can provide improved support. The concepts explained in this section have long years of proven success in World Class wiring harnesses manufacturing operations, and the concepts certainly pass the "Common Sense Test."

If the foreman (or foremen) in your assigned area(s) is not managing by these concepts, you can help him/her realize this fact and help him/her understand and implement the concepts.

Why Worry about Improved Productivity, Mexican Labor Is Cheap (and Philippine)

(Although it is getting more and more expensive every day.)

Low Direct Labor efficiency affects a lot more than Direct Labor cost. In fact, it impacts almost all aspects of our total factory cost.

- Floor space requirements
- Tooling requirements

- Equipment requirements
- Support heads
- Salary heads
- Benefits
- Supplies
- Operating expenses
- Scrap
- Premium freight
- And, so forth

A Direct Labor productivity improvement of 10 points (from 50% efficiency {as was currently being run in both the Philippines and Mexico to an efficiency of 60%}) could reduce the factory cost per hour by up to 14%.

$$\frac{60\% \text{ Eff.} - 50\% \text{ Eff.} \times 70\% \text{ direct labor impact on factory cost}}{50\% \text{ Efficiency}}$$

$$= 14\% \text{ Factory Cost Improvement}$$

Wiring Harness Departmental Organization Basics

Departments should have around 50 Direct operators under the control of a foreman. See Figure 5.2 for a sketch of the way a wiring harness department should be properly organized by operator categories.

- A significantly larger department may need an assistant foreman or group leader.

Each operator must perform a productive job every day (except when training or on special assignment).
Operators and lines must be held accountable for their performances:

- Accurate standards must be established and posted.
- Daily production reports must be completed and recorded.
- Follow-up must be accomplished.

Proper operator assignment is *critical* (as depicted in Figure 5.2 for a final assembly department).

- Visually identify EACH hourly operator and hourly indirect support person.
- Remember that we will get the performance that we motivate.

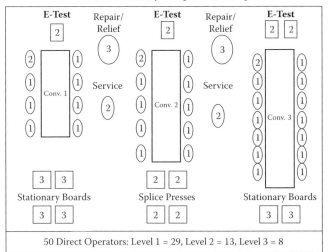

- **Operator Assignments**
    > Assign Level 1 Operators to conveyors, flow lines, U-cells, etc.
    > Assign Level 2 Operators to conveyor inspection stations, E-Test stations, packing, service, splicing, & misc. off-line operations
    > Assign Level 3 Operations to cover absenteeism, repair/relief, stationary boards, critical off-line operations, training, special projects, problem resolution, etc.
- **Use Level 3 Operators to Cover Absenteeism**
    > Supplement with Level 2 Operators as needed
- **Approximate Oprator Percentages by Level**
    > Level 1: 40–60%
    > Level 2: 20–40%
    > Level 3: 10–20%

Figure 5.2   **The importance of providing proper manufacturing support.**

Level 1 Operators

- Assign to Final Assembly Departments on conveyors, flow lines, and U-cells.
    - Simplest, most easily controlled, but most physically demanding, least desirable jobs.
    - Starting place for ALL new operators.
    - Dismiss operators not able or willing to perform.
- Identify best Level 1 operators for advancement to Level 2, but **only when openings exist.**

Level 2 Operators

- Selected from the best of Level 1 operators.
    - It must be a fast and fair process.

- Selected by the Department foreman based on documented performance, attitude, attendance, dependability, flexibility, etc.
- Must make 100% efficiency on their assigned jobs every day or be returned to the line.
  - Final Assembly Departments.
  - Assign to less critical off-line jobs, such as splicing, twisted pairs, etc.
  - Assign to Electrical Test Stations.
  - Assign to the last conveyor station (inspection station).
  - Assign to the service function.
- Assign to standard Lead Prep Department processes (sit-down presses, splices, etc., and to service).
- Identify best Level 2 operators for advancement to level 3, but **only when openings exist.**

Level 3 Operators

- Selected from the best of Level 2 operators.
  - It must be a fast and fair process.
  - Selected by the foreman or foremen and general foreman/supervisor.
  - Select based on documented performance, attendance, attitude, training ability, communication skills, people skills, flexibility, upward mobility, etc.
- Final Assembly Departments.
  - Assign to stationary boards, critical off-line operations, and to off-line operations where excess capacity exists.
  - Use to cover absenteeism.
  - Equalize on the conveyor and rotate on other jobs.
  - Must know and be able to make the standard on all conveyor stations and all other jobs within the department.
  - Responsible to get lines started on time (in support of the foreman).
- Lead Prep Departments.
  - Must know and be able to make the established standards for each of the lead prep processes.
  - Assign to the most critical lead prep jobs.
  - Use to cover absenteeism.
  - Provide setup support.
- Cutting Departments.
  - Assign as cutter operators.

- Accomplish normal setup.
- Use to provide training.
- Assign to special projects, problem resolution, and continuous improvement at the discretion of the supervisor.
- Identify the best of Level 3 for advancement to Level 4 when openings exist, or provide opportunities to move laterally into a support area when openings exist with an opportunity for promotion after a year.

Level 4 Operators

- Selected from the best of Level 3 operators.
  - It must be a fair and reasonably responsive process.
  - Should be a plant-wide selection process with H.R. involvement.
  - Should be selected based on leadership qualities, documented performance, attendance, attitude, dependability, people skills, training ability, communication skills, upward mobility, etc.
- A Final Assembly or Lead Prep Department will generally **not** need to assign Level 4 operators (or assistant foremen/group leaders) except in the cases of very large departments with special challenges.
- Final Assembly (if applicable, normally it will not be).
  - Assist foreman as directed in problem resolution, continuous improvement, special projects, and other assignments as given. (The recommended productive assignment for group leaders is Repair/Relief, when assigned.)
- Lead Prep (if applicable, normally it will not be).
  - Assist foreman as directed on problem resolution, continuous improvement, special projects, and other assignments as given.
- Cutting.
  - Cover absenteeism.
  - Provide cutter expertise and training.
  - Support approximately four cutter operators by providing major setup support and by servicing wire, terminals, and seals.
  - Provide staggered lunch support to increase up-time.
  - Help manage cutting sequencing and inventories.
- Identify the best of Level 4 operators for advancement when openings exist.
  - For assistant foreman or foreman.
  - For support area technician/clerk/analyst/etc.

Foreman's Responsibilities and Support Required

Recognize that being a manufacturing supervisor is a challenging, difficult, high-pressure job, and, as such, it is imperative that the proper support be provided.

- Must provide well engineered manufacturing systems.
  - Conveyors where applicable.
  - Good methods, material flow, layouts, and standards for all processes.
- Provide timely problem resolution.
  - Weekly Priority Meetings coordinated by IE.
  - On-the-floor support from all support areas.
- Quick reaction from HR on people issues.
- Elimination of material downtime and quality issues.
- Production Control must level schedules to the extent possible.
- Eliminate nonvalue-added assignments/paperwork to the maximum extent possible.
  - The greatest savings to the company can be realized through productivity/efficiency improvements on the manufacturing floor.
- Provide a meaningful manufacturing control system.
  - IE to provide accurate standards and a Direct Labor Bible (DLB), also called a routing for each product.
  - System to provide accurate operator and departmental efficiencies.

Manufacturing Supervisor Qualifications

- Good people skills and leadership ability
- Ability to handle pressure and remain calm in a crisis
- Good analytical and problem-solving skills
- Ability to be tough and demanding in a diplomatic way
  - Must have high expectations for employees and for support organizations
- Possess a high energy level
- Well organized and able to develop and maintain records
- Have an understanding of labor standards, operator loading, efficiencies, etc.

Job Responsibilities

- Be in the department 30 minutes prior to start of shift to ensure everything is in order.

- Read notes from prior shift (the first shift foreman is considered the lead foreman).
- Create a Hot Sheet for any materials that will not last the day.
- Initiate corrective action for any other department issues/problems.
- Review schedule requirements.
- Ensure lines are set up to run the correct part numbers.
- Greet operators as they arrive.
- Ensure Level 3 operators are prepared to get lines started on time.
  - Equalization Book is used to determine who goes on the line first.
- Assign off-line Level 2 operators based on schedule requirements. These assignments should be rotated to develop and enhance skills and to prepare the best Level 2 operators for future Level 3 opportunities.
  - Level 3 operators can be assigned to assist with this function.
- Assist with assignment of Level 3 operators.
  - They will often self assign based on equalization and rotational agreements.
- Follow up on departmental issues/problems and get resolutions.
  - Involve Level 3 operators at his/her discretion.

Time Breakdown

- Spend approximately 60% of the day managing by walking around.
  - Talk to operators.
  - Look for problems and improvement opportunities.
  - Ensure that the lines are starting and stopping on time and that department employees are busily engaged on their assignments.
  - Talk to other foremen to optimize operator department assignments, identify operators for potential advancement, share ideas/best practices, etc.
  - Work with engineers and other personnel to resolve problems and implement continuous improvement.
- Spend 20% of the time planning the job.
  - Operator assignments.
  - Department scheduling.
  - Engineering change implementation.
  - Special projects, continuous improvement, cost reduction, etc.
  - Talk to operators in a more structured format to give positive feedback and recognition, train, give constructive counseling, gain information, etc.

- Spend 20% of the time on necessary documentation.
  - Develop and review operator performance data; identify opportunity areas and work with operators where needed.
  - Develop and give employee evaluations.
  - Develop and review department performance data; Identify opportunity areas and develop plans to achieve improvements.
  - Develop notes for next shift supervisor.

Note: Notice that no time was mentioned for meetings and training programs. A supervisor should virtually never be pulled out of his/her department during the department's hours of operation unless someone is sent to manage the department in his/her absence (general foreman, Manufacturing manager, etc.), unless the department is so well organized and running so smoothly that the foreman can be absent for a short period of time without worry.

Assistant Foremen/Group Leaders

- In typical departments (50+/− direct heads), an assistant foreman/group leader should not be necessary, assuming the department is organized properly and the Level 3 operators are doing what they are supposed to do. Sometimes, it will be appropriate to have a single department with 70 or more direct heads (it doesn't make sense to split a high-volume/high work content package into more than one department), which is more that a single foreman can effectively manage. In this case, an assistant foreman or a group leader should be assigned, who will take direction from the lead foreman and will perform many of the mundane tasks so that the lead foreman can spend most of his/her time working with the people.
- If group leaders are used in Final Assembly, it is recommended that they be given assignments that will "force" them to do a required amount of direct labor, such as packaging, repair/relief, or service.

Additional Advice

- A supervisor should understand that he/she must treat all employees fairly; however, this does not mean that all employees must be treated the same. For example, it is not acceptable to allow one Universal operator to accomplish only 90% efficiency on a regular basis, but demand 100% from all the rest. However, if an outstanding employee needs a few days off to take care of some urgent family business, you may give them an entirely different answer than you would give to a very marginal employee who made the same request.

- **My 5–90–5 Rule** must be understood. This rule states that 5% of the employees within a department are so conscientious that no matter how they are managed they will do a great job; 5% of the employees will do a lousy job, no matter how they are managed. And, most importantly, the remaining 90% will flow to the left or to the right, depending on how the 5% groups are managed. Remember, "people will perform as they are motivated to perform." **If you reward good performance and punish bad performance, the giant middle majority will do a good job.**
- Ensure that each operator level (as well as hourly "support people" from departments other than manufacturing) can be visually identified (badge color, smock color, etc.). Let the designations of level be symbols of pride and recognition. Build the desire in lower level operators to achieve the next level and beyond. Managers walking through the plant also will quickly become aware of individuals in places they are not supposed to be (although visually identifying workers will tend to keep them in the right places).
- When dealing with "problem" operators, your level of patience should be related to the type of problem. If an operator has a good attitude and is trying hard, but is having trouble making a known fair standard, I would recommend giving him/her every opportunity to succeed. Sometimes, this type of operator, when he/she finally gets the hang of it, will become an excellent long-term employee who will be very satisfied to work at Level 1 jobs for his/her career. However, if operators are not able to do the job after being given the benefit of the doubt, they must be let go. On the other hand, if the problem is related to poor attitude, lousy attendance, or some other cancerous problem, the patience of the supervisor should be very short.
- It is imperative that each and every Level 2 and Level 3 operator achieves 100% efficiency for each hour he/she is assigned to a productive job with a known fair standard. If not, after being given a fair opportunity to correct the problem, he/she must be returned to the line.

Proposed Hourly Employee Classification Matrix for Wiring Harness Manufacturing

Level	Position Titles	Characteristics/ Promotional Criteria	Wages
1	Paced Operation Operator (*Conveyor, Turntable, etc.*) Flow Line Operator (*U-cell, CMM, etc.*) General Assembler (*Components, etc.*)	All new hires, Average Operators, and "Problem" Operators	Minimum Wage
2	Service Operator Final Assembly Off-Line Operator E-Test Operator Conveyor Inspection Operator Lead Prep Operator	Documented Performance, Attitude, Attendance, Dependability, Flexibility	Level 1 Wages + ____%
3	Universal Final Assembly Operator Universal Lead Prep Operator Cutting Operator Special Build Operator Material Handler (*Warehouse*) Quality Auditor **Methods Lab Operator**	Same as Level 2 plus Training Ability, Communication Skills, People Skills	Level 2 Wages + ____%
4	Universal Cutting Operator Team Leader Sr. Quality Auditor **Sr. Methods Operator** Maintenance/Process Technician	Same as Level 3 plus Leadership Skills, Upward Mobility	Level 3 Wages + ____%
5	Assistant Foreman **Methods Technician** Quality Technician **Sr. Maintenance/Process Technician**	Same as above plus potential for salary position	Range starting at Level 4 Wages + ____%

Note: Other support departments also may have hourly positions that should be in harmony with this matrix.

Chapter 6

When You Measure Performance, Performance Improves

In the last chapter, I explained that the average productivity in a Mexican Final Assembly Department was about half of that achieved within a typical Warren Final Assembly Department (at least, in comparison to the Warren Final Assembly Departments of the 1970s). As I noted, about 80% of this gap was due to the improper organization and management of the Mexican Final Assembly Departments and due to the elimination of the tools used in Warren to control direct labor efficiency.

One of the first training sessions I received in the IE Training Program when I started with Packard in 1971 was on routings, which I have renamed Direct Labor Bibles (DLBs). A routing was defined as a documented master plan for building a product. It detailed each element of direct labor required to build that product, with the appropriate labor standards and allowances for each work element. DLBs were written for each product by the IE who had responsibility for the area in which the product would be built. These DLBs were the foundation for the productivity (efficiency) control system, which was very effectively used by Packard and provided a tremendous benefit in improving manufacturing productivity.

Before I go any further, I should probably touch briefly on "direct labor" since there is probably not a universal agreement on just what this term entails and how it is differentiated from "indirect labor." The definition that most people at Packard would give is that if an hourly person's efforts are directly applied to the manufacture of a product then it is direct labor. But, service operators, who only service materials to the line and do not actually build anything, were considered direct labor, as were repair operators, relief operators, and even some maintenance support, and it is certainly questionable if these functions fit neatly within this definition. The definition I prefer to use is that direct labor is any hourly labor involved in the manufacturing process which can be easily defined and for which standards can readily be established for inclusion on a DLB.

The bottom line is that if you are going to control hourly labor, either direct or indirect, it must be done in a defined system. If it is not done in a labor efficiency (productivity) system, it must be done within the plant operating report (plant budget) or some similar system. For example, the last operator on each Packard conveyor line is an inspector who does minimal production while inspecting the harness as it is being removed from the build tool and then folds it and packs it in the shipping container, but they were included in the DLB, as they should have been. The plant auditors, which floated from department to department, performing audits on selected products, were considered indirect labor and their time was paid for in the plant operating report as a percentage of standard hours earned and shipped within the plant (although it certainly was possible, and might have been better, to include this function on the DLB as a percentage of the labor standards for each of the operations included on the DLB that needed to be audited). As long as the total hours used by auditors was no more than the hours generated on the plant operating report, the plant was within its budget for this function.

From my experience, it is much easier to control hourly labor which is identified and included on a DLB and in an efficiency (productivity) system than it is in a plant operating report (plant budget), so my default position would always be to include hourly labor in the DLB, unless it is just not practical to do so.

From Routings to Direct Labor Bibles (DLBs)

I'm not sure how and when Packard coined the term "routing." I presume it was in the early days, apparently before everyone who was working at Packard in 1971, when I started with Packard, had begun their careers; and I presume the document started as a listing of every piece of material and every direct labor operation that was required to build a product and how each was "routed" through the facility. I doubt that the labor standards, which were part of the system when I began my career, were part of the original documents.

However, I do not think that the word "routing" conveys what I think this document should convey. Because I changed the old Packard system in many significant ways, including the use of computer technology and the implementation of several new concepts and applications, I much prefer the term Direct Labor Bible or DLB for the document I will describe and detail in this chapter, which captures all of the direct labor required to build any product in the plant, and which is the foundation for the Computerized Direct Labor Productivity System. A system such as this is critical to control direct labor and to reduce direct labor costs.

The standards utilized on the DLBs were established in one of three primary ways. The first way was through the use of Standard Data. This was primarily used for cutting and lead prep operations where processing was very consistent and straightforward. When writing the DLB, the IE would reference the Standard Data Tables and select the appropriate standard based on the operation being performed and the specifications of the lead or assembly.

The second way was through the use of a reference standard. Reference standards were used mainly for establishing standards for relatively simple splices or subassemblies. When the IE was writing a DLB and needed to establish a standard in one of these cases, he would look in the back of the Standard Data Book and try to find a similar assembly. If he could, he would use that standard for the new process or make modifications based on estimated differences.

The third process was preplanning, which was the most time consuming, but it was also the most important and valuable, because it was used to establish accurate standards (as well as detailed methods, layouts, and quality specs) for conveyor lines, stationary boards, and all other processes for which good standards were not available. (Preplanning is the subject of the next chapter.) Once the preplanning process was complete, the accurate standard was available to be input on the routing.

Every final assembly finished product that was built in the entire operation had its own DLB. Depending on the size of the harness, it could take anywhere from a couple of hours to more than a day to write a single DLB, but the time spent was well worth it. (DLBs for additional part numbers in the same package were written much more quickly, as only the differences needed to be incorporated.)

Warren DLBs were far from perfect. The unit of time used for standards was "hundredths of an hour" (called "points"). It would certainly have been easier to use and understand (or visualize) minutes per piece than points per piece.

Another fallacy of the system was that average allowable delays were included in each of the standards rather than minimum allowable delays. An allowable delay is defined as any nonproductive time observed and allowed within a process that had been identified based on a series of studies, such things as paid breaks, setups, machine downtime, service interruptions, supervisor instruction, gauging and inspection, etc. (Nonallowed time was not included in the standards.)

At the end of this chapter is a document (DLB Philosophy) that explains in detail why this is a problem. However, the bottom line is that when most operators achieve 100% of a standard rate for a given time period, they will normally stop production, even if they are experiencing a very good day with minimum delays and they should actually be able to produce more. When they are experiencing a day with average delays, they will generally produce the standard in the time available, but, on days when they are experiencing above-average delays, they will not be able to meet the standard (unless they work at above 100% effectivity), and the production they lose will not be made up on good days. The bottom line is that they do not bank production on good days to offset the losses on bad days.

Of course, the other problem with Packard DLBs was that they were manually written and took a long time to complete. When we started up the Mexican Operations, I submitted a project to enhance the DLB and efficiency system by fixing the two problems mentioned above, along with other structural improvements. No dice, I would have to wait until I got to South Korea where I would be able to revamp the entire system the way it should be done using the latest technology.

In spite of these problems with the DLBs, the fundamental DLB and efficiency control systems were sound and enormously helpful to manufacturing supervision in controlling their departments and plants, as well as being very important tools in the costing and pricing systems.

Production Efficiency

The Packard DLB system was used to generate two different types of efficiencies. The first type Packard called an External Efficiency, but which I will refer to as Production Efficiency It is a very accurate measure of a department's productivity based on shipped hours. Each DLB, one per finished final assembly part number, contained not only the individual labor standards for each of the operations that would be required to produce the product, it also contained the total labor, in points per harness, which would be paid to each of the departments that performed any of this labor. In most cases, there would only be three departments that would be included on a DLB for a given part number: the Cutting Department, the primary Lead Prep Department, and the Final Assembly Department. Sometimes there were other departments included, for example, those departments that had the processing capability for specialty assemblies that were not available in the primary Cutting or Lead Prep Departments.

Whenever harnesses were completed, placed in shipping containers, and physically shipped to the warehouse, the various departments were paid (assigned standard hours) for the work they had done to help produce these products. For example, let's suppose that a Chevrolet Impala Engine harness had a DLB value of 30.000 minutes (for simplicity, I will use minutes instead of points) for Final Assembly, 10.000 minutes for Lead Prep, and 5.000 minutes for Cutting. If 1000 harnesses were shipped from Final Assembly Department A on a given day, Final Assembly Department A would earn 500.0 standard hours (1000 harnesses × 30.000 min. per harness/60 minutes per hour), the Lead Prep Department would earn 166.7 standard hours, and the Cutting Department would earn 83.3 standard hours.

The computer system being used was quite archaic, so the Production Efficiency Report was only run on a weekly basis and on the last day of the month. What the report provided was the total number of standard hours generated for each department based on the harnesses that had been shipped to the warehouse and how much each department was paid on the DLB for the work it had done. Each Final Assembly Department would generate hours for harnesses that were built and shipped from the department: Lead Prep Departments would earn hours from harness built and shipped from the three or four Final Assembly Departments that they supplied, and Cutting Departments (because they were centralized for a plant) would generate hours from harnesses built and shipped from all Final Assembly Departments within the entire plant.

In addition to the total number of standard hours earned by each department, the report also included the total number of clock hours that were used within each department during the concurrent time period, based on timecard data. The number of standard hours generated was divided by the number of clock hours used to establish a Production Efficiency (Department Efficiency based on product shipped).

The Production Efficiency was quite accurate and provided a very good reflection of the actual performance of each department. All of the standards on the DLBs were quite accurate and they were consistent with the standards on the DLBs for other departments, so the various department efficiencies were quite comparable, i.e., Final Assembly to Final Assembly, Lead Prep to Lead Prep, and Cutting to Cutting. The end of the month numbers certainly provided the motivation to continue to try to improve performance, because the plant managers were looking at these numbers with extreme interest, largely because the managing director was also looking at them, along with a host of other people.

Some people felt the system could be abused and could lead to unhealthy competition. However, the only significant way the system could be abused would be for the IE to manipulate the DLB, and that just didn't happen; there was no upside for an IE to do so. Besides, the Budget DLB, which was the initial labor estimate (based on Standard Data, Reference Data, or estimates) that was utilized in the pricing process, served as a check and balance. I also never witnessed unhealthy competition; good natured ribbing, perhaps, but not unhealthy competition. The system kept everyone in manufacturing on their toes, which was a good thing.

However, even though the Production Efficiency was a fair refection of performance, it had a major failing—it didn't provide information that would allow a foreman, supervisor, or manager to pinpoint specific problems and initiate corrective action. The department Production Efficiencies, while accurate, could not be broken down by operator, or even by type of operation. This is why the second efficiency report was critical. It did provide the detailed data necessary to identify and solve specific problems (at least, to a point).

Process and Operator Efficiencies

The second type of calculated efficiency was called an Internal Efficiency by Packard but which I will refer to as Process Efficiencies and Operation Efficiencies (P&O). This tool was based on the concept that "when you measure performance, performance improves." An Internal Efficiency Report was generated on a daily basis to determine the efficiency of each operator and each different type of process within each Final Assembly and Lead Prep Department. (I'm not sure about Cutting, but I don't think they calculated Internal Efficiencies, even though it would have been very beneficial to do so.) However, this report was not generated by the computer, it was generated by hand. A "desk girl" was assigned to each general foreman's area, comprised of a Lead Prep Department and three or four Final Assembly Departments.

At the end of each day, the foreman would fill out a "rate card" for each of the conveyors, which provided the quantity of each part number built on each of the conveyor lines during the shift, along with a total number of clock hours utilized on the line to generate this production. Also, every off-line operator turned in a daily rate card. The rate card for the Service operators usually indicated that they had worked the entire day on servicing. The rate card for the Repair operator and the Relief operator indicated how many hours they had spent on their primary function, along with other productive work they had done within the department. The expectation was that they would meet the standard (100% efficiency) on other jobs they did outside of their main functions.

Each Universal operator, who was not filling in for absenteeism on a conveyor line or doing a special project for the foreman, was assigned a productive off-line job. Their rate cards (which they filled out) would indicate the jobs they did, the quantity they produced, and the number of hours they were engaged on the activity. The clear understanding was that each of them would achieve 100% efficiency on the productive jobs they did during the shift (jobs and/or functions captured in the DLBs).

The desk girl would take the rate cards and detail them on a one-page form. Sometime the following day, the foreman would receive this form, which detailed the prior day's production data. By individual and by type of process, the hours earned and the hours utilized would be shown, along with a summary of the entire department's performance. Because these reports were done by hand, there was no record maintained, other than these daily forms. I designed my own spreadsheet on which I could put a month's worth of data, and I recorded the performance of each conveyor and each Universal operator, as well as the Relief operator and the Repair operator. Additionally, I keep a running total of the overall hours earned within the department versus the hours used. There is no question that keeping this spreadsheet helped me control and improve the performance within my department.

In a well-managed Final Assembly Department, the conveyors would make the established rate every day, so the hours earned would be equal to the clock hours utilized. A good foreman usually made a few hours per shift in servicing, because he or she could calculate the number of service hours that would be earned each shift (assuming the standards would be made on all operations) and he or she would round down to the nearest whole service person. A good foreman also would earn a few hours in repair, as he or she would keep defects below the Packard norm. The same thing goes for emergency relief, as a good foreman would control emergency relief such that the Relief operator had time to generate several productive hours per day. Each of the Universal operators would achieve 100% efficiency per day on whatever off-line job they were assigned, almost as a given.

Considering the above information, a good foreman could earn a few extra standard hours on a good day. He or she would put these into the bank, because invariably something would happen during the month that would cost standard hours, such as power outages, gigantic snow storms that kept a lot of people from getting to work, material outages, etc. In addition, sometimes Universal operators would be assigned to special projects that did not generate standard hours; assignments, such as special cleanups, sorting of contaminated materials, etc.

Wouldn't One Efficiency Number Be Sufficient?

Both efficiencies were very important because they provided different information, and they also provided a check and balance. P&O Efficiencies were needed to specify performance by an individual operator and by the type of operation in order to allow for detailed analysis and corrective action. The P&O Efficiency was a combination of individual Operator Efficiencies (in which individual operators generated standard hours, e.g., stationary boards, off-line splices, service, etc.) and Process Efficiencies where more than one operator was responsible for generating the standard hours, e.g., conveyors. Production Efficiencies did not provide that capability.

Production Efficiencies were highly accurate and dependable, because these efficiencies were based on actual harnesses that arrived in the warehouse. While P&O Efficiencies were actually more useful to a foreman in helping him or her run a productive department, they were less reliable;

The Production Efficiency provided the necessary check and balance. There are only three things that could cause the P&O Efficiency to diverge from the External Efficiency:

1. P&O Efficiencies were calculated on daily production for each process, whereas Production Efficiencies were calculated based on harnesses shipped out of the department to the warehouse. If there was a buildup of inventory during the month, the P&O Efficiency would be higher than the Production Efficiency. If there was a draw-down of inventory during the month, the opposite would be the case. However, over time, inventories would be relatively stable, hence, this was generally not much of a factor.
2. Any material that was scrapped after it left the workstation was reported as production and was included in the P&O Efficiency, but it was not included in the Production Efficiency because the scrapped material never made it into a finished harness that was shipped out of the department. This was normally a very small percentage; usually less than 1%.
3. The last possible factor was falsification of rate cards. If an operator overstated his or her production, it would inflate the Operator Efficiency, but would not generate standard hours (obviously) on the Production Efficiency Report. Falsification of documents was a very serious violation of the work rules and punishable up to and including termination (although that rarely happened). By comparing the P&O Efficiency to the Production Efficiency, a foreman could ensure that the rate cards he or she was receiving were pure. The fact that everyone knew this was being done helped keep them that way.

The Warren Bottom Line

With the utilization of these efficiency control (productivity control) tools, along with understanding how to properly organize and manage manufacturing departments, good foremen in Lead Prep and Final Assembly Departments in Warren would achieve 100% production efficiency on a month in/month out, year in/year out basis. The exception would be the month of a major model change. A big model change might reduce the production efficiency of a well-managed department to the mid-80% the first month, the production efficiency would be back into the mid-90% the second, and the department would be back to running 100% production efficiency the third.

This was typical performance in Warren in the 1970s. Many departments, in each of the harness building plants, would consistently run production efficiencies at or near 100%, based on fair standards for each of the products they produced. Overall plant production efficiencies would run in the mid-90% and higher. Now, all that Packard had to do was to replicate this performance in Mexico (and everywhere else it intended to set up shop) or not.

So, What Did We Do in Mexico?

When Packard started the first plant in Mexico, all of the DLBs were in place for all of the products and Production Efficiencies were calculated. I wanted to fix some of the inherent problems with the system, as I explained, but I was unsuccessful in getting MIS (Management Information Systems, i.e., the computer department) approval. But, at least the system we had used in Warren, with a great deal of success, was in place for this control.

However, as hard as I tried, I could not convince the Manufacturing manager to institute the P&O Efficiency System, which, if anything, was more important than the Production Efficiency System. He had never worked as a first line supervisor (foreman) nor a second line supervisor (general foreman) for that matter, so I just couldn't convince him of the importance. After all, Mexican labor is cheap.

During the first couple of years in Mexico, when there was still a lot of excitement and enthusiasm, our average efficiency was 35% to 40% lower than Warren's. A good chunk of this was due to the lack of a system to control the productivity of the operators within the various departments, which was just the thing P&O Efficiency would do. A few years later, when the enthusiasm had waned, and after Packard had introduced remote cutting and lead prep and had terminated the DLB and Production Efficiency System, the Mexican production efficiencies were over 50% less than what they had been in Warren in the 1970s.

But, hey, Mexican labor is cheap (right?) even though it was by far the biggest line item cost, as it is for virtually any manufacturing company in virtually any corner of the world. This 50% decrease in production efficiency or productivity versus what was typically accomplished in Warren, only cost the company several tens of millions of dollars annually.

I Finally Got a Chance to Do It Right in South Korea

I don't intend to talk much about my experience in Portugal regarding efficiency (productivity) control, because what I found in our European Operations was a situation that was worse than what had existed in Warren. There was nothing resembling a P&O Efficiency, and the Production Efficiency System did not provide a tool with much value. The DLBs (called ZMGs) that were written throughout Europe were unique to each plant, so, it was not possible to use the Production Efficiencies for any type of benchmarking or comparative purposes. This never made any sense to me, although the plant managers seemed to like the system.

After I had the Portuguese Methods Lab up and running and all of the engineers trained, I spent some time designing an automated Productivity System, which included both P&O and Production Efficiencies, which could be used universally throughout Packard's Operations. Because operators from around the globe are equally capable of building wiring harnesses, everything else being equal, there would be no differences in labor standards. The only differences within the system were in the various delays and allowances, which could vary by location due to such things as local agreements (for such things as paid breaks, time allowed at the start and end of shift for preparation, etc.), dependability of electrical power, availability of skilled technicians, etc. This system would make efficiencies comparable around the globe and would provide a tool for each region, and for every operation within that region, to pinpoint specific problem areas and help them implement corrective actions.

Shortly before I was asked to go to South Korea as the Joint Representative director of the new joint venture that was to be initiated, I presented my proposal for revamping the entire "DLB and Efficiency System" (Automated Productivity System) to a large group in Germany that included the European Engineering director and most of the European plant managers. The Engineering director was extremely interested in the system I proposed, especially since it was a system that would provide comparable data for each of the plants in Europe, regardless of whether that plant was in Portugal, Ireland, England, Germany, or anywhere else.

However, it was apparent that most of the plant managers were not nearly as enthusiastic.

A few weeks later, I was asked to take the Joint Representative director's job in South Korea, which I accepted. I expected that I would be asked by the Engineering director to provide information on the system about which he had seemed so excited, and help them get started on the implementation before I left. The request never came.

However, I knew this would be one of the very first systems that would be implemented in South Korea, and it was. If I remember correctly, this system was the second system we put into place, which followed the Computerized Cutting System. I believe the Material Control System was the third.

By 1986, computers had come a long way from where they were in 1971. I took full advantage of this in designing and then implementing the Computerized DLB and Efficiency System (Productivity System) in South Korea. In the early 1970s, it would take a day or more to write a DLB for a single Instrument Panel Harness. It would take another hour or more to write a DLB for each additional part number in the package. Utilizing the new computerized system, DLBs for the entire package could be written in an hour or two, depending on the number of lead prep operations.

What was so time consuming in producing a manual DLB was the requirement to determine on which cutter type each lead would be cut, then having to look up the standard for each lead in the standard data book, and finally having to manually record this data on the DLB. For an Instrument Panel Harness with several hundred leads, this took quite a while. With my computerized system, this was done automatically based on the circuit information, which had already been input into the Computerized Cutting System, matched with the database information for each type of cutter. All that had to be done in order to generate a DLB was to input the lead prep information, which would take less than an hour for even the largest harness, and the final assembly information, which would take only a few minutes. There was also a Budget DLB, in addition to this Production DLB, which was identical except that estimates were used for those processes that were not available in Standard Data and for which preplanning had not yet been accomplished, e.g., the labor required to build the harness on the conveyor. This Budget DLB was used in the Costing and Pricing System before the preplanning process even started.

When You Measure Performance, Performance Improves

Once all of the lead prep and the final assembly processes had been input, the DLB could be printed out in a couple of minutes and, depending on which format was selected, there was room for at least three additional part numbers within the package on the same document.

In addition to the Computerized Cutting System, this system pulled data from two other systems already in operation: the Time Clock System, which provided data on hours worked by operator by department, and the Material Control System, which provided data on harnesses shipped from each department to the warehouse.

In order to print out all of the reports I had designed (most especially the P&O and the Production Efficiency Reports), the following data had to be input on a daily basis:

- Daily production data for the conveyors

Line # (1): 7 Board Code: 09C25
Description: 2009 Chrysler Power Harness
Department: 105 Shift: AM
Date: 5/11/09

Conveyor Daily Production Report

Station #	Operator #	Operator Name	Substitute Operator (2)	Additional Operator (3)	Minutes Paid for conv Work (4)	Operation Code (5)	Part Number (6)	Pieces Produced (6)	DLB Min. per Piece (7)	DLB Minutes Earned (8)	Down-time Code (12)	Down-time Min. (9)	# of Oper. Down (10)	Minutes on Non-DLB Activities (11)	Remarks
1	1004	Ima Humpin			530	FA8-01	87654321	195	24.500	4777.5	UC03	5.0	10	50.0	
2	1790	I.M. Runnen			530										
3	1530	I.B. Builden			530										
4	589	R. U. Certin			530										
5	478	Uni Versal	X		530										
6	423	Ura Hurtin			530										
7	1579	I.B. Taipen			530										
8	398	I.R. Sloe			530										
9	7	I.M.Waiten			530										
10	549	I. Inspekt			530										
8	492	Flex Abel		X	300										Sta. 8 gets behind
					(A) 5600					(B) 4777.5				[C] 50.0	

85.3%	Conveyor labor efficiency =	4777.5	(B) DLB Minutes earned /	5600	(A) Total minutes paid		
86.1%	Net conv. labor efficiency =	4777.5	(B) DLB Min.earned /(	5600	(A) Total minutes paid	- 50.0)	[C] Min. lost on non-paid activity

1. Only one report is to be turned in for the entire conveyor line by the Foreman or Department Technician.
2. If the line operator is substituting for the regular operator due to absenteeism or turnover, put a mark (X) in this column.
3. Do not list operators that are training new operators (they will turn in an individual Daily Production Report); list only extra operators that are assigned to "help-out" on the line.
4. Indicate the number of clock minutes each operator is being paid to work on the conveyor (including paid breaks).
5. Note the code for final assembly conveyor build from the DLB. This will allow DLB standard minutes/piece and DLB std. piece/hour to be automatically accessed.
6. List each part number built on the line during the shift and the quantity of pieces taken off of the line per part number.
7. Indicate the DLB standard minutes per piece (Plant Standard) and the DLB standard pieces per hour for each part number built during the shift. This information is found on the DLB or provided by the I.E.
8. DLB min. earned = pcs. produced (6) x std. per piece from the DLB (7) summed for all part numbers produced.
11. Minutes spent on non-paid activity (i.e., not included in the DLB) = # of downtime minutes (9) x # of operators affected (10).
12. Uncontrollable Downtime Codes: UC01 - Raw materials not in plant; UC02 - Training programs, seminars, etc.; UC03 - Loss of power; UC04 - Engineering change rework; UC05 - Sorting, unavoidable rework; UC06 - Other uncontrollable downtime
 Controllable Downtime Codes: C01 - Plant responsible downtime; C02 - Department meetings; C03 - Major maintenance; C04 - Repair of department caused defects; C05 - Training new operator; C06 - Special cleanup; C07 - Other controllable downtime
 If the automated Productivity System is in place, all of the highlighted blocks are automatically calculated.

174 ■ *Intelligent Manufacturing: Reviving U.S. Manufacturing*

■ Daily production data for all off-line operators

Operator Daily Production Report

Operator (1)	Ima Splicer		Operator #	1396
Department	103		Shift	AM
Date	5/10/09		Clock Minutes (2)	530	(A)

Board Code	Operation Code (3)	Part Number (4)	Operation Description (5)	Minutes Worked on Paid Operations (6)	Pieces Prod.	DLB Standard Minutes per Piece (7)	DLB Std. Pcs. Per Hour (7)	DLB Standard Minutes Earned (8)	Net Eff. % (8)	Down-time Code	Minutes Spent on Non-Paid Activity (10)	Remarks
09x01	FA3-03	12345678	140ABCDE Splice	350	500	0.520	115.4	260.0	74.3%	UC02	50	Excellence Training
		12345678	150ABC Splice	130	500	0.220	272.7	110.0	84.6%			
				480				370.0			50	
(11)				[C]			(D)		(B) Minutes lost on non-paid activities			

[C] Minutes worked on paid operations	=	530
Gross Operator efficiency	=	370.0
Net operator efficiency	=	370.0

480
69.8%
77.1%

(A) Clock minutes — 530 (A)Clock minutes
(D) Standard minutes earned / 480 [C] Minutes worked on paid operations
(D) Standard minutes earned /

1. At the end of each day, each operator that is assigned to any off-line job is to fill out and turn in a daily production report. This information will be used to establish individual operator efficiencies and identify opportunity areas. (*This includes all operators not assigned to a conveyor station, whether the job they are assigned to is paid for through a DLB or not. Non-paid jobs would include such things as training, special department clean-up, sorting of material, attending a meeting, etc.*)
   The operator will input all of the information onto the form except the shadowed columns and the calculations at the bottom (*these will be automatically calculated*).
2. Clock Minutes (the minutes that the operator is paid each day) includes paid breaks but does not include the non-paid lunch (*a paid lunch would be included*).
3. The operation code for each job that an operator performs will come from the DLB.
4. Indicate the harness part number for each specific operation that is listed. The base part number can be listed if the operation is common to all (*or many*) part #s in a package.
5. The description should be reasonably detailed (*140ABC splice, 1SA and 1SB twisted pair, etc*) so that it will be easy to clarify any questions/errors that might arise.
6. Paid Operations are operations which are included in the DLB. The operator should record the number of minutes worked on each operation, including job prep and clean-up, setup, self service, breaks and personal time.
7. This information should be provided to the operator by the Foreman (or IE) or accessed directly from the DLB.
8. These calculations are automatically made.
   DLB standard minutes earned = Σ (pieces produced × DLB standard minutes per piece).
   Net Efficiency % = DLB standard minutes earned on each operation ÷ minutes worked on each operation
9. List the code for any reportable downtime.
10. This is the number of minutes of downtime for the code indicated.
11. The Minutes Worked on Paid Operations (included in a DLB) and the Minutes Spent on Non-paid Operations must be equal to the Clock Minutes. If there is a discrepancy, it should be resolved with the operator. Adjust the Minutes Worked on paid or non-paid activities as necessary, so that the minutes add-up correctly.
    \> The gross operator efficiency is based on total DLB standard minutes (or hours) earned versus the clock minutes (or hours) paid to the operator. Therefore, any downtime negatively affects the gross operator efficiency.
    \> The net operator efficiency is based on DLB standard minutes (or hours) earned versus the actual minutes worked to generate those standard minutes (or hours), excluding Controllable and Uncontrollable Downtime. Therefore, downtime does not negatively affect net operator efficiency.

Note 1: Job prep and clean-up, setup, self service, 1st piece inspection, breaks, and other customary elements of the job are included in minutes (hours) worked since they are paid in the Base Standard or they are incorporated in the Base Standard Adjustment.

Note 2: If the Automated Productivity System is in place all of the highlighted blocks are automatically calculated.

When You Measure Performance, Performance Improves ■ 175

■ Data on the temporary transfer of operators between departments

Direct Operator Time Transfer Sheet

Loaning Department _____
Supervisor _____
Supervisor _____

Personnel _____
Page _____ of _____

Date	Shift	Employee Name	Employee #	Receiving Department	Time to be Transferred		Reason / Job Done
					Hours	Minutes	

Receiving Department	Hours to Receive	Supervisor	Signature	Date	Remarks

■ Data on uncontrollable and controllable downtime

Downtime Sheet

Plant _____

Date ___/___/___
Time _____
Page ___ of ___

This is an official request by _____, Supervisor of department _____, to remove the following non-productive (*lost*) hours from the net efficiency calculation for the department.

Lost Time Code	Number of Operators Affected	Total Time Lost		Clarifications / Explanations / Justifications / Remarks
		Hours	Minutes	

Controllable Downtime

Code	Description
C01	Plant Responsible Material Downtime
C02	Department Meetings
C03	Major Maintenance
C04	Major Rework Due to Department Error
C05	Training New Operators
C06	Special Clean-up
C07	Other Controllable Downtime

Uncontrollable Downtime

Code	Description
UC01	Raw Material Not in Plant
UC02	Training Programs, Seminars, etc.
UC03	Loss of Power
UC04	Engineering Change Rework
UC05	Sorting, Unavoidable Rework
UC06	Other Non-Controllable Downtime

Mfg. Supervisor _____

Approved _____ ___/___
Approved _____ ___/___

It only took a few hours, by one clerk, to input this data on a daily basis for a plant of 500 hourly operators. With this input, the system allowed me to design reports that provided me, as well as my manufacturing manager and supervisors, with all of the information needed in order to identify problems and implement corrective actions. Just the knowledge that this system existed provided a lot of motivation to the operators to do a good job. And, it provided a lot of motivation to manufacturing supervision to run efficient operations, evidenced by the fact that the first year this system was in place, our efficiency was 20 points higher than Mexican efficiency.

I could ask for a report any day of the week (although I generally had my reports delivered on Mondays), which would give me the Production Efficiencies for each department on a daily, month-to-date, and year-to-date basis; the Process Efficiencies by type of operation for each department and for the plant; the Operator Efficiencies for each individual operator (who had done any off-line work) sorted in any manner desired (by department, from best to worst, by type of operation); a downtime summary by department or plant and by category of downtime; a detail of improvement by part number; and about anything else I wanted to know regarding productivity.

After I reviewed the reports, I would spend time with my managers and supervisors to review things I found "interesting." I made sure to emphasize all of the positives, especially improvements that had been made. However, when it was clear that problems existed, it was important to discuss these issues. There is no question that just having the various reports available had a very positive impact on our performance, but then utilizing the reports effectively accelerated our improvement.

I probably spent no more than a couple of weeks contemplating and designing this system, and the programming took no more than a couple of months including debug time. But, the system undoubtedly saved us millions of dollars annually.

How Much Could the Mexican Operations Have Saved?

After returning to the States, I made efforts on several occasions to try to convince Packard's upper management to restore Production Routings (DLBs) and External (Production) Efficiency Reports by department. All that remained from a once superior system (albeit, one lacking in technological

Production Efficiency Report (and Improvement Summary)

Date ___/___/___
Plant _____

Dept. #	Dept. Type	Clock Hours (Daily, MTD, YTD)	Standard Hours Earned from DLB	Production Efficiency % (Daily, MTD, YTD)	Budget Hours Earned from Labor Estimate	Budget Efficiency %	Budget Improvement Hours	% Budget Impv.	Plant Improvement Hours	This as a % of Manufactured Hours Earned	Hours Paid But Not Worked	This as a % of Clock Hours	Uncontrollable Downtime Hours	This as a % of Clock Hours	Net Clock Hours	Net Production Efficiency	Net Budget Efficiency
101	Cutting	xxx	xxx	XX%	xxx	XX%	xx	X%	xx	X%	xx	X%	xx	X%	xxx	XX%	XX%
		xxxx	xxxx	XX%	xxxx	XX%	xxx	X%	xxx	X%	xxx	X%	xxx	X%	xxxx	XX%	XX%
		xxxxx	xxxxx	XX%	xxxxx	XX%	xxxx	X%	xxxx	X%	xxxx	X%	xxxx	X%	xxxxx	XX%	XX%
102	Lead Prep	xxx	xxx	XX%	xxx	XX%	xx	X%	xx	X%	xx	X%	xx	X%	xxx	XX%	XX%
		xxxx	xxxx	XX%	xxxx	XX%	xxx	X%	xxx	X%	xxx	X%	xxx	X%	xxxx	XX%	XX%
		xxxxx	xxxxx	XX%	xxxxx	XX%	xxxx	X%	xxxx	X%	xxxx	X%	xxxx	X%	xxxxx	XX%	XX%
103	Final Assembly	xxx	xxx	XX%	xxx	XX%	xx	X%	xx	X%	xx	X%	xx	X%	xxx	XX%	XX%
		xxxx	xxxx	XX%	xxxx	XX%	xxx	X%	xxx	X%	xxx	X%	xxx	X%	xxxx	XX%	XX%
		xxxxx	xxxxx	XX%	xxxxx	XX%	xxxx	X%	xxxx	X%	xxxx	X%	xxxx	X%	xxxxx	XX%	XX%
104	Final Assembly	xxx	xxx	XX%	xxx	XX%	xx	X%	xx	X%	xx	X%	xx	X%	xxx	XX%	XX%
		xxxx	xxxx	XX%	xxxx	XX%	xxx	X%	xxx	X%	xxx	X%	xxx	X%	xxxx	XX%	XX%
		xxxxx	xxxxx	XX%	xxxxx	XX%	xxxx	X%	xxxx	X%	xxxx	X%	xxxx	X%	xxxxx	XX%	XX%
105	Lead Prep	xxx	xxx	XX%	xxx	XX%	xx	X%	xx	X%	xx	X%	xx	X%	xxx	XX%	XX%
		xxxx	xxxx	XX%	xxxx	XX%	xxx	X%	xxx	X%	xxx	X%	xxx	X%	xxxx	XX%	XX%
		xxxxx	xxxxx	XX%	xxxxx	XX%	xxxx	X%	xxxx	X%	xxxx	X%	xxxx	X%	xxxxx	XX%	XX%
106	Final Assembly	xxx	xxx	XX%	xxx	XX%	xx	X%	xx	X%	xx	X%	xx	X%	xxx	XX%	XX%
		xxxx	xxxx	XX%	xxxx	XX%	xxx	X%	xxx	X%	xxx	X%	xxx	X%	xxxx	XX%	XX%
		xxxxx	xxxxx	XX%	xxxxx	XX%	xxxx	X%	xxxx	X%	xxxx	X%	xxxx	X%	xxxxx	XX%	XX%
107	Final Assembly	xxx	xxx	XX%	xxx	XX%	xx	X%	xx	X%	xx	X%	xx	X%	xxx	XX%	XX%
		xxxx	xxxx	XX%	xxxx	XX%	xxx	X%	xxx	X%	xxx	X%	xxx	X%	xxxx	XX%	XX%
		xxxxx	xxxxx	XX%	xxxxx	XX%	xxxx	X%	xxxx	X%	xxxx	X%	xxxx	X%	xxxxx	XX%	XX%
	Total Plant	xxx	xxx	XX%	xxxx	XX%	xxx	X%	xxx	X%	xxx	X%	xxx	X%	xxxx	X%	XX%
		xxxx	xxxx	XX%	xxxxx	XX%	xxxx	X%	xxxx	X%	xxxx	X%	xxxx	X%	xxxxx	X%	XX%
		xxxxx	xxxxx	XX%	xxxxxx	XX%	xxxxx	X%	xxxxx	X%	xxxxx	X%	xxxxx	X%	xxxxxx	X%	XX%

Budget Improvement Hours are Budget Hours Earned from the labor estimate (Budget DLB) minus Standard Hours Earned (from the DLB) based on product shipped from the Final Assembly Department, and Budget Improvement % equals Budget Improvement Hours divided by Budget Hours.

There is tremendous flexibility to design reports as you want them. In addition to Production Efficiencies and Budget Efficiencies, this report shows Net Production Efficiencies and Net Budget Efficiencies; which are calculated by deducting Uncontrollable Downtime and Hours Paid But Not Worked (which is sometimes mandated by law in Mexico and other locations) from Clock Hours. It also shows Plant Improvement Hours in order to get a feel for improvements being made on the floor, over and above the Budget Improvement, which is primarily generated through the preplan process.

This entire report is computer generated, and while all of the information in this report is helpful, by far the most important data is the highlighted column which is the Production Efficiency including all delays and downtime included. This column really tells us how each of the departments and the plant is running.

When You Measure Performance, Performance Improves ■ 179

Plant _____
Department _____

Specific Process Efficiency Report

Date ___/___/___

Period	Shift	Description	Operation Code (Process) 1	Operation Code (Process) 2	Operation Code (Process) ....N	Service	Rework	Department Allowance (1)	Low Volume Allowance (5)	Part Number Proliferation Allowance (5)	Plant Improvement (5)	Downtime Hours (2)	Hours Not Reported (3)	Net Transfer Hours (4)	Total Department by Shift	Total Department for All Shifts Combined
Day __/__	A	Clock Hrs	XXX	XXX	XXX	XXX	XXX	XXX				XXX	XXX	XXX	XXX	XXXX
		DLB Hours Earned	XXX	XXX	XXX	XXX	XXX	XXX	XXX	XXX	XXX				XXX	XXXX
		Efficiency %	XX%	XX%	XX%	XX%	XX%	XX%							XX%	XX%
	B	Clock Hrs	XXX	XXX	XXX	XXX	XXX	XXX				XXX		XXX	XXX	
		DLB Hours Earned	XXX	XXX	XXX	XXX	XXX	XXX	XXX	XXX	XXX				XXX	
		Efficiency %	XX%	XX%	XX%	XX%	XX%	XX%							XX%	
	C	Clock Hrs	XXX	XXX	XXX	XXX	XXX	XXX				XXX	XXX	XXX	XXX	
		DLB Hours Earned	XXX	XXX	XXX	XXX	XXX	XXX	XXX	XXX	XXX				XXX	
		Efficiency %	XX%	XX%	XX%	XX%	XX%	XX%							XX%	
MTD __/__	A	Clock Hrs	XXXX	XXXX	XXXX	XXXX	XXXX	XXXX				XXXX	XXXX	XXXX	XXXX	XXXXX
		DLB Hours Earned	XXXX	XXXX	XXXX	XXXX	XXXX	XXXX	XXXX	XXXX	XXXX				XXXX	XXXXX
		Efficiency %	XX%	XX%	XX%	XX%	XX%	XX%							XX%	XX%
	B	Clock Hrs	XXXX	XXXX	XXXX	XXXX	XXXX	XXXX				XXXX	XXXX	XXXX	XXXX	
		DLB Hours Earned	XXXX	XXXX	XXXX	XXXX	XXXX	XXXX	XXXX	XXXX	XXXX				XXXX	
		Efficiency %	XX%	XX%	XX%	XX%	XX%	XX%							XX%	
	C	Clock Hrs	XXXX	XXXX	XXXX	XXXX	XXXX	XXXX				XXXX	XXXX	XXXX	XXXX	
		DLB Hours Earned	XXXX	XXXX	XXXX	XXXX	XXXX	XXXX	XXXX	XXXX	XXXX				XXXX	
		Efficiency %	XX%	XX%	XX%	XX%	XX%	XX%							XX%	
YTD __/__	A	Clock Hrs	XXXXX	XXXXX	XXXXX	XXXXX	XXXXX	XXXXX				XXXXX	XXXXX	XXXXX	XXXXX	XXXXXX
		DLB Hours Earned	XXXXX	XXXXX	XXXXX	XXXXX	XXXXX	XXXXX	XXXXX	XXXXX	XXXXX				XXXXX	XXXXXX
		Efficiency %	XX%	XX%	XX%	XX%	XX%	XX%							XX%	XX%
	B	Clock Hrs	XXXXX	XXXXX	XXXXX	XXXXX	XXXXX	XXXXX				XXXXX	XXXXX	XXXXX	XXXXX	
		DLB Hours Earned	XXXXX	XXXXX	XXXXX	XXXXX	XXXXX	XXXXX	XXXXX	XXXXX	XXXXX				XXXXX	
		Efficiency %	XX%	XX%	XX%	XX%	XX%	XX%							XX%	
	C	Clock Hrs	XXXXX	XXXXX	XXXXX	XXXXX	XXXXX	XXXXX				XXXXX	XXXXX	XXXXX	XXXXX	
		DLB Hours Earned	XXXXX	XXXXX	XXXXX	XXXXX	XXXXX	XXXXX	XXXXX	XXXXX	XXXXX				XXXXX	
		Efficiency %	XX%	XX%	XX%	XX%	XX%	XX%							XX%	

In the first column MTD is Month to Date and YTD is Year to Date.

1). The Clock Hours for "Department Allowance" are the hours utilized on non-routed jobs; such as training, special clean-ups, sorting material, etc.
2). "Downtime Hours" is the sum of Controllable and Uncontrollable Downtime reported on the Operator and the Conveyor Daily Production Reports.
3). "Hours Not Reported" is the difference between actual Clock Hours and the sum of the Clock Hours reported on the Operator and the Conveyor Daily Production Reports. This should be zero; it it is not, find out why.
4). Net Transfer Hours is the difference between the hours of direct labor loaned to other departments and the number of direct hours borrowed from other departments.
5). All of these calculations for Standard Hours Earned (including for the Department Allowance) are based on percentages from the DLB applied to the various process categories.
6). The DLB Hours Earned should be with a percent or so of the DLB hours earned on the Production Efficiency Report. If not, find out why. The Production Efficiency should be very accurate.

Individual Operator Efficiency Report

Department _____
Shift _____
Date ___/___/___

Month-to-Date | Year-to-Date

Employee Number	Name	Operation Code	Operation (Process) Description	Clock Hours	Not Reported Hours as % of Clock Hours	Reported Hours by Process Type	Std. Hours Earned by Process Type	Efficiency % vs. Reported Hours	Efficiency % vs. Clock Hours	Downtime hours as % of Clock Hours	Clock Hours	Not Reported Hours as % of Clock Hours	Reported Hours by Process Type	Standard Hours Earned by Process Type	Efficiency % vs. Reported Hours	Efficiency % vs. Clock Hours	Downtime hours as % of Clock Hours
1000	xxxxxxxxxxxxxxxxx	xxxxx	xxxxxxxxxxxxxxxxxxx			XXXX	XXXXX	XX%					XXXXX	XXXXX	XX%		
		xxxxx	xxxxxxxxxxxxxxxxxxx			XXXX	XXXXX	XX%					XXXXX	XXXXX	XX%		
		xxxxx	xxxxxxxxxxxxxxxxxxx			XXXX	XXXXX	XX%					XXXXX	XXXXX	XX%		
	Operator Total			XXXX	X%	XXXX	XXXXX	XX%	XX%	X%	XXXXX	X%	XXXXX	XXXXX	XX%	XX%	X%
1001	xxxxxxxxxxxxxxxxx	xxxxx	xxxxxxxxxxxxxxxxxxx			XXXX	XXXXX	XX%					XXXXX	XXXXX	XX%		
		xxxxx	xxxxxxxxxxxxxxxxxxx			XXXX	XXXXX	XX%					XXXXX	XXXXX	XX%		
	Operator Total			XXXX	X%	XXXX	XXXXX	XX%	XX%	X%	XXXXX	X%	XXXXX	XXXXX	XX%	XX%	X%
1002	xxxxxxxxxxxxxxxxx	xxxxx	xxxxxxxxxxxxxxxxxxx			XXXX	XXXXX	XX%					XXXXX	XXXXX	XX%		
		xxxxx	xxxxxxxxxxxxxxxxxxx			XXXX	XXXXX	XX%					XXXXX	XXXXX	XX%		
	Operator Total			XXXX	X%	XXXX	XXXXX	XX%	XX%	X%	XXXXX	X%	XXXXX	XXXXX	XX%	XX%	X%
1003	xxxxxxxxxxxxxxxxx	xxxxx	xxxxxxxxxxxxxxxxxxx			XXXX	XXXXX	XX%					XXXXX	XXXXX	XX%		
		xxxxx	xxxxxxxxxxxxxxxxxxx			XXXX	XXXXX	XX%					XXXXX	XXXXX	XX%		
	Operator Total			XXXX	X%	XXXX	XXXXX	XX%	XX%	X%	XXXXX	X%	XXXXX	XXXXX	XX%	XX%	X%

This report could be set up to provide additional detail, if desired; for example, it could contain Net Clock Hours, in addition to clock hours, by deducting hours assigned to non-DLB jobs and then calculating a Net Operator Efficiency; etc.

The Operation Code is the generic process code from the DLB; i.e., the code for splicing, for stationary board assembly, etc., and not the specific code for the splice or stationary board assembly.

Not Reported Hours (the difference between actual clock hours and reported clock hours) should be zero. Make sure each operator knows that a Daily Production report must be submitted.

The Downtime Hours as a % of Clock Hours is based on the total of the downtime hours reported by the operator on the Daily Production Reports. If these percentages are not in reasonable balance among Universal Operators, find out why. Controllabe and uncontrollable downtime hours could be split, if desired, and Net Efficiency calculations could be made - there is a lot of flexibility to design reports to give you exactly the information you want, in the format you want.

The most important numbers on this report are the Operator Efficiencies versus Clock Hours which are highlighted. These numbers indicate the **true productivity** of each operator.

updating) was a Budget Routing (DLB) and a plant-wide Budget Efficiency—both almost worthless for the purposes of benchmarking or for driving improvement. Even more importantly, I tried to convince Packard to reintroduce Internal Operator and Process Efficiencies, which would have been invaluable in helping control direct labor, which by then was largely out of control.

Several times, I made formal presentations in high-level meetings about this system that we had so effectively used in South Korea, and while there were always several individuals who seemed very interested in introducing such a system, no formal approvals to get the money to make it happen were ever given.

Finally, about six months before I left Packard (and about 10 years after I had left South Korea), I made another presentation to the Mexico West Operations staff regarding the Computerized Productivity System I had developed. I started the presentation by asking each of the managers (representing each plant and each of the functional areas) to estimate the improvement in direct labor efficiency that we could expect to experience within each plant if a series of tools was provided, which we currently did not have. (These tools, not surprisingly, were the tools that would be supplied by the Computerized Productivity System.)

There were about 10 managers in the meeting, and the lowest estimated productivity (efficiency) savings was 10%, the highest was 40%, and the average was over 20% (which was about what most of the plant managers estimated). I told them that for the Mexico West Operations, a 20% improvement in Direct Labor would result in a cost savings of over $30 million dollars per year, over $50 million per year savings if Mexico East was included. I then suggested that if they did not agree to fund this project, it was for one of two reasons: either they did not believe the improvement estimates that they had just made, or they didn't think it made sense to spend a couple of hundred thousand dollars—one time—in order to save $50 million every single year.

Finally, it looked like I had found the right approach and we got the green light to proceed. I was assigned some engineering talent to get started accumulating and fine-tuning standard data and to establish proper delays and allowances. It even looked like a couple of programmers were going to be assigned. Then, the 2001 budget crunch hit, and the plug was pulled on this program, a program that would cost a couple hundred thousand dollars to put into place and would save tens of millions of dollars annually. For me, this was the last straw. I made an immediate decision to get out while the getting was good at the first feasible opportunity.

Here is the case study example I used in this meeting:

Productivity Case Study

You are the Foremen for a Final Assembly Department of about 50 employees. Currently, you have no information on your department's productivity performance other than what you can deduce through observation. You are meeting the schedules you are given, have good quality results, & try to see that people are busy; but it is very difficult to keep an eye on everyone at all times considering everything you have to do. You know that each department needs to reduce costs & that the biggest element of cost, by far, within the operation is direct labor; but you don't know where to start.

You finally decide that if you are going to make any real headway in your department, you are going to need to ask Industrial Engineering for help with tools that can help you control direct labor within your department. They promise to provide all the help you need, but only on the condition that you commit to a specific productivity improvement at the end of a six-month period.

What will be your productivity improvement commitment if Industrial Engineering agrees to provide you with the following tools? :

TOOLS

1. Accurate standards and a detailed report that will allow you to forecast and assign operators; as well as an accurate efficiency report that will allow you to monitor your department's **overall** performance on a daily, weekly, monthly, and YTD basis & compare it to the performance of your colleagues' departments.   (_____% Improvement)

2. Tool #1, with the addition of a report that will allow you to monitor the labor efficiency (daily, weekly, monthly, and YTD) on each type of process within your department (such as splice presses, stationary boards, conveyors, off-line ring-out stations, service, etc.); as well as a report that details all downtime in the department.   (_____% Improvement)

3. Tools 1 & 2, with the addition of a report that will allow you to monitor the labor efficiency of every operator within your department on a daily, weekly, monthly, & year to date basis.   (_____% Improvement)

As a Plant Manager, what overall plant improvement commitment would you make if each department had level 3 tools & you had a report that provided detailed performance data by department and plant, which would provide an accurate comparison of your plant with other plants in the Mexican Operations?   (_____% Improvement)

Note: The average improvement estimated by the Mexico West Staff, which would be realized in the Mexico West Operations if this entire system were put into place, was 20%. A 20% labor improvement for Mexico West equated to about $30 million per year. For all of Mexico, it equated to about $50 million annually. The cost to implement the system was a couple hundred thousand dollars, including all programming time and the engineering effort required to establish all of the necessary standard data. When budgets were cut, this system was the first casualty - it should have been the last.

When You Measure Performance

There is just absolutely no question that "when you measure performance, performance improves." If you have any doubts, just imagine professional sports if no scores were kept, or imagine our school system if no grades were given (actually, the last example is being tried in a few places, with the kind of results you would expect).

For sizeable companies, I think a computerized system, something like the one I have talked about, with both Production Efficiencies and with Operator and Process Efficiencies, makes the most sense. Let the computer do all of the hard work of maintaining, calculating, sorting, and distributing performance information, which is critical in controlling the most important and the most costly resource within virtually any manufacturing company. But, whatever you do, find some way to measure and record performance, even if it means using a manual spreadsheet to record the actual outputs versus the expected outputs of each operator.

W. Edwards Deming said that work standards or quotas should be eliminated on the shop floor. He was right about a lot of things, but he was dead wrong about this. Just look at Packard's Mexican Operations performance if you want confirmation. When workers know that their performance is being recorded and reviewed (and it doesn't have to be, nor should it be, done in a threatening way), both the quantity and quality of their output will be positively impacted.

I have included a few pages from the IE Manual in the remainder of this chapter that I wrote about the DLB and the Computertized Productivity System, which should be helpful in understanding some of the concepts reviewed up to this point in this chapter. Once standard data is in the database for the various processes (including process standards, Base Standard Adjustments, Department Allowances, Service Allowances, Repair Allowances, and Low Volume Allowances and Part Number Proliferation Allowances where applicable, all of which will be explained in the remainder of this chapter), reports can be designed to provide virtually any information desired on productivity, headcount requirements, tooling and equipment requirements, labor improvement, etc.

**WHAT YOU DO
NOT MEASURE YOU DO
NOT CONTROL !!**

**WHEN YOU
MEASURE PERFORMANCE,
PERFORMANCE IMPROVES !!**

**WHEN YOU MEASURE &
REPORT PERFORMANCE,
PERFORMANCE IMPROVEMENT
ACCELERATES !!**

Packard Productivity Nose Dives

The graph below represents the degradation of Packard's productivity, starting in the late 1970s and through the end of the century. It highlights some of the key factors responsible for this dramatic decline, which already have all been touched on in this book. It is an extremely disappointing graph, especially for a company that at one time was world class in the control of direct labor.

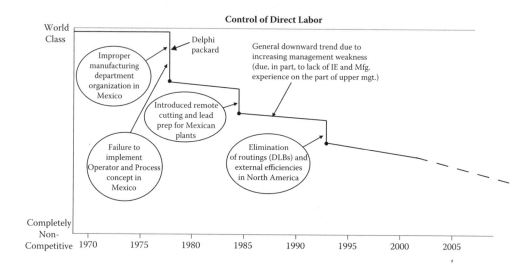

Are We Really Controlling Productivity in Our Plants?

1. Do Manufacturing supervisors need an accurate document to help them in the planning, organizing, and assigning of personnel within their departments in order to maximize the utilization of human resources?
2. To most effectively manage our human resources, is it important to understand the productivity within our facilities, not only on an overall plant basis, but also down to the specific department, conveyor line, type of operation, and **individual operator?**
3. Is it important that a manager be able to **objectively** compare the productivity of one operator versus another operator, one department versus another department, or one plant versus another plant, in order to assess performance and create healthy competition?
4. Is it important that each operator has a fair standard against which his/her performance is measured, such that when he/she has a good day

(i.e., encounters minimum delays) he/she can achieve 100% attainment of that standard assuming he/she performs at 100% effectivity and works precisely the number of hours being compensated?
5. Is it important that management is able to understand the true cost of manufacturing a product versus the estimated cost, understanding that the estimated cost was used to establish a price to the customer?
6. Is it important that management has a tool that will accurately forecast the actual cost to produce a product so that initial pricing to customers is neither too high (meaning the business will probably not be won) nor too low (meaning that the business probably will be won, but the company will lose money in the process)?

If the answer to most, if not all, of the above questions is yes, is there a system (or systems) in place to accomplish these issues? If not, what would it be worth to have such a system?

Direct Labor Bibles (DLBs)

A DLB (similar to what Packard once called a Routing, but with significant enhancements) is defined as a documented master plan for building a product, which details each element of direct labor required to build that product with the appropriate labor standard for each work element. There are actually two kinds of DLBs: Budget DLB and Production DLB.

The **Budget DLB** (or Labor Estimate) is the estimate of direct labor established for a product during the estimating process. This labor estimate establishes the budget labor for the product, which will be used to establish a price (in conjunction with the fully burdened company cost per budget hour and the material content). The budget labor totals will be displayed on the DLB/Budget/Improvement Report (along with the Production DLB labor totals and the Improvement/Deprovement).

Note: The Budget DLB should be revised based on the introduction of any engineering change, either prior to or during production, which has an effect on labor, except that once a product is in production, the budget should not be adjusted downward, regardless of the change, if it is not possible to actually achieve the labor reduction (e.g., if the engineering change would only result in additional line balance).

Note: Budget DLBs should be as reflective of the actual direct labor required to build a product as possible. If a fat Budget DLB (or estimate) is established, the direct labor content will be overstated and the company may price itself out of the market.

Note: On the other hand, an unrealistically low Budget DLB means that the direct labor content of the harness will be understated, which will reduce the company's profit margin for that product, or even create a loss. If this happens on enough products, it won't take a company long to be in severe financial difficulties.

The **Production DLB** is the document that details the actual direct labor content of a finished final assembly product when it goes into production. It may be identical to the Budget DLB, but it will generally vary somewhat, depending on the introduction of engineering improvements or differences between estimated standards (where no standard data exist) and the actual standards that are established through Methods Lab studies (preplanning), master studies, etc.

The total DLB should be no higher than, and hopefully less than, the total estimated direct labor. This difference between the two direct labor totals for a product is called budget improvement (or deprovement, if the budget labor total is less than the Production DLB labor total). An improvement means that the company will increase the profit margin on the product (assuming that something has not happened to drive the fully burdened labor rate higher) from what was initially projected. Meanwhile a deprovement means that the company will not meet the projected profit margin for that product, or may even lose money.

Efficiencies from a Budget Routing

Plant Productivity Information Available Based on Labor Estimates (Budget DLBs) Only

- Plant Budget Efficiency history can be used to provide some benchmarking within a plant, but only if the business within that plant remains fairly constant.
- Plant Budget Efficiency numbers are generally *not* comparable from plant to plant (or from department to department, assuming that individual departments have been established within a plant).

Detail of Efficiency Information Based on Budget Routings

PLANT   X Y Z

65% BUDGET EFFICIENCY

Since Budget Efficiencies are based on Budget Routings, which include estimated standards (when no standard data exist), the resulting efficiencies are rarely comparable between departments or plants.

- Plant Budget Efficiency data do *not* provide a meaningful management tool to:
  - Determine and assign headcount.
  - Provide meaningful comparisons between plants, departments, operations, or operators.
  - Identify specific problem areas within departments, by type of process, or with specific operators.
  - Determine the effectiveness of corrective actions and improvement initiatives, which have been implemented.
  - Control individual operator performance.

Plant Productivity Information Available with an Automated Productivity System Based on Production DLBs

- Provides consistent Production process, and Operator Efficiency data from plant to plant, department to department, operation to operation, and operator to operator.
  - Based on actual work measurement or accurate standard data.
  - Incorporates adjustments for low volume and part number proliferation.
  - Incorporates allowances and delays based on extensive studies.
- Provides employee motivation.
  - Standards can be achieved on days with minimum delays.
  - Performance can be accurately compared and benchmarked.
- Provides an outstanding management tool.
  - Thorough, yet simple, reports provide easy management review of realistic performance.
  - Accurate planning tool for headcount determination, tooling and equipment requirements, and floor space requirements.
  - Provides a tool to determine the effectiveness of Lean initiatives.
  - Provides a tool to validate cost reduction projects.
  - Provides on the floor control to:
    - Identify department organizational problems.
  - Identify problem operations.

Plant Productivity Information Available When Production, Process, and Operator Efficiencies Are Based on Direct Labor Bibles (DLBs)

Plant X Y Z = 60% Production Efficiency

DEPT. 1: 70% Production Efficiency

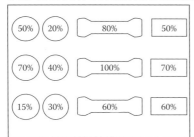

DEPT. 2

60% Production Efficiency

Process & Operator Efficiency data is also available for this department, similar to Dept. 1

DEPT. 3

50% Production Efficiency

Process and Operator Efficiency data is also available for this department, similar to Dept. 1

DEPT. 4

65% Production Efficiency

Process and Operator Efficiency data is also available for this department, similar to Dept. 1

DEPT. 5

45% Production Efficiency

Process and Operator Efficiency data is also available for this department, similar to Dept. 1

DEPT. 6

60% Production Efficiency

Process and Operator Efficiency data is also available for this department, similar to Dept. 1

ADVANTAGES
- Effective management tool
- Improves focus on specific problems
- Provides employee motivation
- Provides a consistent and accurate performance measurement tool throughout the operations

REQUIREMENTS
- Accurate DLBs (Direct Labor Bibles)
- Computerized Efficiency System (Production, Operator, and Process)
- Discipline on Daily Production Reports
- Knowledgeable, committed supervision

– Identify problem operators.
– Determine effectiveness of corrective action.

Terms and Definitions Used with the Computerized Productivity System

- **Base Standard**: This is the actual time required to perform a complete operation, including allowed nonstandards, but excluding all delays. In the case of machine controlled cycles, the Base Standard is simply the cycle time of the machine.
- **Base Standard Adjustment (BSA)**: This adjustment is the percentage of time during which no production occurs due to allowed constant delays (breaks, specified job prep and clean-up, etc.) and the *minimum* allowed variable delays (job setup, service interruptions, machine downtime for material problems or maintenance, etc.), which have been observed in a series of Delay Studies. Simply stated, the BSA represents the minimum percentage of lost productive time for a specific type of operation for allowed delays.
- **Plant Standard** (or Attainable Standard): This is the standard (in minutes/piece) that is obtained by increasing the Base Standard by the BSA percentage. The Plant Standard (in min/pc) = the Base Standard (in min/pc) × (100% + the BSA percentage). The Plant Standard (in pieces/hour) = 60 minutes/hour ÷ the Plant Standard (in min/pc). This represents the maximum attainable output by an operator who is working at a 100% effectivity level and who is having a good day (i.e., experiencing minimum delays).
  - Generally, the operator will fall short of achieving the Plant Standard (in pieces/hour) because only the minimum allowable delays are included in this standard, but on a good day they should attain the standard.
- **Service Standard**: This is the direct labor required to move and handle all materials and products within a production department, including the proper placing of all required materials into production workstations. Service also includes general housekeeping within a production department.
  - Service is paid in the DLB as a percentage of all of the direct labor subtotals within each department (at rates established through Delay Studies), which can be found in Standard Data. If service is not included in the DLB, it must be included in the Plant Budget (or Plant Operating Report) or controlled and costed by some other means.
- **Repair/Scrap Standard**: This is the direct labor required to repair defective assemblies/products produced within a production department

or the labor used to produce the assemblies/products if the material is to be scrapped. Repair is included in the DLB as a percentage of all direct labor subtotals within each department at rates established through engineering studies. The repair/scrap rates can be found in Standard Data.
- It should be noted that repair required due to engineering changes or missing and/or defective raw material is not covered by the standard. This cost should be charged to the customer and/or supplier.
- **Department Allowance (DA)**: This is the difference between the minimum allowed delays (BSA) and the average allowed delays, expressed as a percentage of productive time. The percentages allowed for each operation category are found in Standard Data. Theoretically, the Department Allowance will allow the department to achieve 100% production efficiency, assuming all operators within the department are assigned jobs included in the DLB and are working at 100% effectivity while experiencing normal delays.
- **Plant Improvement (PI)**: This is defined as legitimate labor improvement generated through engineering, manufacturing, quality, and/or other support area creativity, innovation, and efforts, after a product has gone into production. This is not to be confused with Budget Improvement, which is the difference between the Budget DLB (or Estimate) and the Production DLB, which will reflect Methods Lab improvements, differences between estimated standards and applied standards, and any "paper improvement."
  - If legitimate improvements are made, such that Plant Standards can be reduced on the Production DLB, then Plant Improvement has been generated. Plant Improvement (in minutes/piece) is the difference between the old Plant Standard(s) and the new Plant Standard(s).
  - This PI will be included on the routing and will remain for one year from the date of introduction. This will give the Production foreman incentive to work with the engineers to create labor improvements, which will offset early losses in efficiency that normally accompany the introduction of new methods, processes, etc. Eventually, PI will result in significant savings.
  - Note that Budget Improvement also will result from Plant Improvement, since service, repair, and allowances are established as a percentage of Plant Standard subtotals, which will be reduced as a result of Plant Improvement.

- **Low Volume and Part Number Proliferation Allowances**:
  These two allowances are paid to the department to offset the negative impact of low volume and/or a high number of part numbers in a package. These allowances are necessary because both of these factors lead to lower productivity (and higher cost) due to increased setups, increased training, new part number startup efficiency losses, increased defects, etc.
    - These allowances are found in Standard Data. If they accurately reflect the true impact of volume and part number proliferation, these allowances will offset the impact of these two factors and we will be able to accurately establish the true productivity level of each department (allowing us to do realistic benchmarking and department comparisons).
    - Note that Routings, along with the accompanying Efficiency Report(s) and other performance reports, are extremely important documents, because they allow a company to define and control direct labor, which, for most manufacturing companies, is far and away the biggest cost item for its products. DLBs are used directly or indirectly to make estimates, establish prices, forecast headcount requirements, determine actual cost, and determine direct labor performance, so it is extremely important that they are kept accurate and up to date.
- Fortunately, automated routings are fairly simple to generate and maintain, and any of the reports can be requested and obtained as desired/required.

Simple DLB Example

Let's suppose that we are gong to introduce a new product into our manufacturing plant called a widget. Because it's a new product, we do not have any data concerning standards, delays, or allowances. Therefore, in order to develop a realistic labor estimate, and ultimately a production DLB, we have decided to establish a complete set of standard data for the assembly of a widget.

To assemble a widget, three different operations are required: metal cutting (three pieces each), subassembly (two assemblies each), and final widget assembly. It will be necessary to establish standard data for each operation.

Productivity System Terms, Calculations, & Philosophy

Base Standard = raw operation time or process time (*established through Master Study, Lab Study, cycle time*)

Base Standard Adjustment (*BSA*) = minimum allowable delays (*established through Delay Studies*)

Plant Standard (*or Operator Standard*) = Base Standard increased by the BSA

Departmental Allowance = Average Delays - BSA

Example: for an operation with a 60 second cycle (*Base Standard*) and a paid work day of 8 hours (*480 minutes*), a total-of 6 delay studies provided the following data:

MINIMUM ALLOWABLE DELAYS	AVERAGE ALLOWABLE DELAYS	ACTIVITY
20 minutes	20 minutes	Paid breaks
5 minutes	5 minutes	Job prep & cleanup
5 minutes	15 minutes	Misc. delays (*meetings, machine downtime, set-ups, self service, supervisor instruction, emergency relief, gage & inspect, etc.*)
30 minutes	40 minutes	

Base Standard Adjustment (*BSA*) = $\frac{30 \text{ min.}}{480 \text{ min.} - 30 \text{ min.}}$ = 0.067 or **6.7%**

Department Allowance = $\frac{40 \text{ min.}}{480 \text{ min.} - 40 \text{ min.}}$ - 0.067 = 0.024 or **2.4%**

Base Standard = 60 sec. / pc. = **1.000 min. / pc.** (60 pcs. / hr.)

Plant Standard = 1.000 min. / pc. X (1 + 0.067 BSA) = **1.067 min. / pc.** (56.2 pcs. / hr.)

Department Allowance = 1.000 min. / pc. X 2.4% = **0.024 min. / pc.**

60.0 pcs./hr. ---------------------------- **Base Standard line** (standard with no delays included)

56.2 pcs./hr. ---------------------------- **Plant Standard line** (attainable output on days with minimum delays)

55.0 pcs./hr. ---------------------------- **Average Output line** (*paid to department in routing*)

Actual operator performance (*assumes operator working at 100% effectivity*)

① Average Output = 60 min./hr. / (1.067 + .024) min./pc. = 55.0 pcs./hr.

Production which is above the Average Output line is banked on good days (when less than average delays are experienced) to offset days when production is below the line due to higher than average delay days.

For this product, there are no applicable Low Volume or Part Number Proliferation Allowances.

Metal Cutting

- An engineer took delay studies of the Metal Cutting Operation for five days and obtained the following data:
  - The automatic cutter operates on a 6.0 second cycle.

- The minimum allowed delay time during any of the five days (unavoidable time that the machine was shut down and not running) was 60 minutes.
  - 30 min: Operator breaks
  - 10 min: Job preparation and cleanup
  - 5 min: Stock change
  - 5 min: Machine adjustment
  - 10 min: Paperwork and miscellaneous
  - 60 min: Total
- The average allowable delay time during the five 8-hour days was 90 minutes.
  - 30 min: Operator break
  - 10 min: Job prep and cleanup
  - 10 min: Stock change
  - 10 min: Machine adjustment
  - 30 min: Paperwork and misc. (i.e., dept. meetings, talk to foreman, power outage, machine breakdown, training programs, etc.
  - 90 min: Total
- For every 1000 pieces of metal cut, 15 minutes of Service and 7 minutes of Repair were required.

■ The engineer then established the following standard data:
- Base Standard = 6.0 sec/pc ÷ 60 sec/min = 0.100 min/pc
- Plant Standard = Base Standard × (1 + Base Standard Adjustment)

$$BSA = 60 \text{ minutes minimum delays} \div (480 \text{ minutes} - 60 \text{ minutes})$$
$$= 0.143 \text{ (see Note 1 at the end of the chapter)}$$

Plant Standard = 0.100 min/pc × (1 + 0.143) = 0.114 min/pc
(equal to 526.3 pcs/hr)

- Department Allowance = [90 minutes avg. delays ÷ (480 min − 90 min)] − 0.143 BSA = 0.231 − 0.143 = 0.088 = 8.8%
- Service Allowance = 15 min ÷ 1000 pcs ÷ 0.114 min/pc Plant Std. = 0.131 = 13.1%
- Repair Allowance = 7 min ÷ 1000 pcs ÷ 0.114 min/pc Plant Std. = 0.061 = 6.1%

Component Subassembly

- The engineer took delay studies of the Subassembly Operation and obtained the following data:
  - The average time for an operator to make one subassembly (including allowed nonstandard elements) was 0.400 minutes. [(.420 + .380 + .360 + .430 + .440 + .440 + .390 + .370 + .390 + .420) ÷ 10]
  - The minimum allowed delay time observed was 45 minutes.
    - 30 min: Breaks
    - 5 min: Job prep and clean up
    - 5 min: Restocking
    - 5 min: Misc.
    - 45 min: Total
  - The average allowed delay time observed was 60 minutes.
    - 30 min: Breaks
    - 5 min: Job prep and clean up
    - 10 min: Restocking
    - 15 min: Misc. (i.e., paperwork, training programs, department meetings, talk to foreman, etc.)
    - 60 min: Total
  - For every 1000 sub assemblies made, a total of 30 minutes of Service and 5 minutes of Repair were required.
- The engineer established the following Standard Data:
  - Base Standard = 400 min/pc
  - Base Standard Adjustment = 45 min ÷ (480 min − 45 min) = 0.103 (see note 1)

  Plant Standard = .400 minutes/pc × 1.103 = 0.441 min/pc
  (equal to 136.1 pcs/hr)

  - Dept. Allowance = [60 min ÷ (480 min − 60 min)] − 0.103 = 0.040 = 4.0%
  - Service Allowance = 30 min ÷ 1000 pcs ÷ 0.441 min/pc = 0.068 = 6.8%
  - Repair Allowance = 5 min ÷ 1000 pcs ÷ 0.441 min/pc = 0.011 = 1.1%

Final Assembly of a Widget

- The engineer took delay studies of the Final Assembly Operation and obtained the following data:
  - The average time for an operator to assemble one widget, including allowed nonstandard elements, was 2.00 minutes.

- The minimum allowed delay time observed was 37 minutes.
  - 30 min.: Breaks
  - 5 min: Job prep and clean up
  - 2 min: Paperwork and misc.
  - 37 min: Total
- The average allowed delay time observed was 50 minutes.
  - 30 min: Breaks
  - 5 min: Job prep and clean up
  - 15 min: Paperwork and misc.
  - 50 min: Total
- For every 100 assemblies, a total of 20 minutes of Service and 8 minutes of Repair were required.

■ The engineer established the following Standard Data:
- Base Standard = 2.000 min/pc
- Base Standard Adjustment = 37 min ÷ (480 min − 37 min) = 0.084 (see note 1)

Plant Standard = 2.00 min/pc × 1.084 = 2.168 min/pc
(equal to 27.7 pcs/hr)

- Department Allowance = [50 min ÷ (480 min − 50 min)] − 0.084 = 0.032 = 3.2%
- Service Allowance = 20 min ÷ 100 pcs ÷ 2.168 min/pc = 9.2%
- Repair Allowance = 8 min ÷ 100 pcs ÷ 2.168 min/pc = 3.7%

Note 1: The Base Standard Adjustment also can be expressed as a percentage and the Base Standard can be multiplied by 100% plus the≈Base Standard Adjustment percentage to obtain the Plant Standard.

Note 2: This example is based on an 8-hour workday. If a 9-hour day is standard, 540 paid min/day should be used, etc.

The following routing for the complete assembly of a widget can now be written:

When You Measure Performance, Performance Improves ■ **197**

Direct Labor Bible

Part Nr(s).	12345678			DLB Rev. #.	Initial				Page 1 of 1
Product Desc.	2008 Widget			Pay Dept. (s)	101,102,103				Date 5/5/08
Drawing Nr.	12345678	Chg. Nr.	Initial	Engineer	I.M. Smart				Ship Dept. 103

Oper-ation	Description	Comp. Part Nr.	Comp. Part Nr.	Source of Std.	PN - 12345678 Vol/Mo - 100,000					PN - Vol/Mo -				
					Base Std.	Plant Std.	Pcs/Hr	Qty	Min/Piece	Base Std.	Plant Std.	Pcs/Hr	Qty	Min/Piece
	Cutting Department (101)													
C - 1	Cut Metal Strip			MS88-1	.100	.114	526.3	3	.342					
C - A	Cutting Dept. Allowance (8.8%)								.030					
C - S	Cutting Dept. Service Allowance (13.1%)								.045					
C - R	Cutting Dept. Repair Allowance (6.1%)								.021					
C - T	**Cutting Department Total**								**.438**					
	Sub-Assembly Department (102)													
SA - 1	Sub-assemble Components			MS88-2	.400	.441	136.1	2	.882					
SA - A	SA Allowance (4.0%)								.035					
SA - S	SA Service Allowance (6.8%)								.060					
SA - R	SA Repair Allowance (1.1%)								.040					
SA - T	**Sub-assembly Department Total**								**.987**					
	Final Assembly Department (103)													
FA - 1	Widget Final Assy			MS88-3	2.000	2.168	27.7	1	2.168					
FA - A	FA Allowance (3.2%)								.069					
FA - S	FA Service Allowance (9.2%)								.199					
FA - R	FA Repair Allowance (3.7%)								.080					
FA - T	**FA DepartmentTotal**								**2.516**					
Total	**Total Product Direct Labor Standard**								**3.941**					

Chapter 7

Preplanning: The Perfect Tool to Accomplish Toyota's Rule #1

Based on the four-year study by Stephen Spear and Kent Bowen, "Decoding the DNA of the Toyota Production System" (*Harvard Business Review*, Sept.–Oct., 1999), and my own observations when visiting NUMMI, Toyota's Rule #1 is: "All work shall be highly specified as to content, sequence, timing, and outcome." They clearly understand that if they do not do an outstanding job of designing and implementing a robust manufacturing system, there is no hope that the output from that system will be world class. The quality of the engineering that goes into the establishment of a manufacturing system, in large part, is going to determine the levels of productivity, quality, delivery reliability, cost effectiveness, investment control, and process cycle time obtained from that system.

When I started with Packard in 1971, there was a world-class system in place, which was used to develop effective manufacturing systems for each of the wiring harnesses produced within the operations. Because of the high cost of labor in Warren, the system was developed to drive the labor content down to the lowest possible level without placing an unfair burden on the operators or expecting them to work at an effectivity level (a combination of pace and efficiency of movements) greater than 100%. But, this system not only minimized the labor content, it also provided for minimum investment and created a work environment that would be most conducive to good quality, excellent delivery reliability, and optimum process cycle time.

This system was called "preplanning" and it was taught to each industrial engineer (IE) within the company as a major focus of the two-week IE Training Program. When I went through the program, an entire week was spent learning the very important principles and concepts that were part of this system. Because it has been proved that the best way to learn something is by doing it, this is exactly how this part of the training program was structured. A very small harness was selected and, during this week, the entire preplanning process was utilized to develop a proposed manufacturing system. There were some necessary shortcuts made, to accommodate the time available, but the entire process was understood at the end of the week as well as its value to the company.

So, just exactly what is preplanning? I think the best definition I can come up with is that it is a structured system that is used to develop the optimal manufacturing system for any product, which uses Methods Lab simulation as the primary means to develop methods, flows, layouts, and standards.

After I had been introduced to this system, it seemed so logical and beneficial that I assumed that preplanning, in addition to being utilized at Packard, was a common industry practice. I later came to discover that this not only was not a common industry practice, it was not even a common industry practice within the wiring harness industry.

What Is the Common Industry Practice?

When I was transferred to Portugal, I had an opportunity to observe and evaluate the various engineering systems that were in place to help me determine where improvement opportunities existed as well as to benchmark systems that could be successfully introduced to Packard's North American Operations. There were a lot of good things being done in Europe, especially in regards to Process Engineering responsibilities, but, their level of IE expertise, which was a big strength at Packard, was quite primitive. I was amazed at the process used to establish manufacturing systems for wiring harness production.

In Packard Electric, an industrial engineer and methods operator team were given a preplan assignment, which contained one or more harness packages to be preplanned. The length of the actual assignment was usually 20 to 24 weeks, which was the approximate amount of time available from the point when sufficiently accurate data were available to start the preplan process until the preplanning needed to be completed so that preparations could be made for start of production. Depending on the harness package size (approximate seconds of work content) and complexity (number of part numbers, introduction of new materials, complicated building techniques needed, etc.), the preplan schedule could be anywhere from 6 to 24 weeks for a given package. The engineer and operator team would use a well-defined and controlled multistep process to establish the most effective manufacturing system possible, using simulation throughout the process to approximate actual plant conditions.

By way of contrast, in Europe an industrial engineer would develop a manufacturing system sitting at his desk. Build fixtures (boards) were designed and built by the board builders, through use of a sample harness, with little, if any, engineering analysis or evaluations. Production plans (number of lines, number of operators per line, line speeds, and station breakdowns) were established through the use of standard data or estimates (which can be very inaccurate for wiring harness final assembly). The good news was that this process took less than half the time of the preplan process used in Packard's North American Operations, and no methods operator was required. However, the bad news was that the resulting product (the labor content of the production plan developed) was 10 to 30% less optimum (higher) than the results obtained through the preplan process (depending on package complexity and the skill level of the IE).

In addition to having significantly higher labor content, conveyor lines, which had not been properly preplanned, historically experienced much more severe problems at production startup; problems such as trying to keep the line running (especially as they approached the goal line speeds), poor quality coming off of the lines, difficulty meeting delivery requirements, etc. In the best of circumstances, production startups for wiring harness conveyors are challenging and some difficulties will be encountered, but, with a good preplan job, these problems are minimized. However, without a good preplan job, production startups can be, and usually are, nightmares.

After we installed a Methods Lab in Portugal and implemented the concepts of preplanning, we experienced dramatic improvements in the production plans that were installed in the plants. In addition, we experienced significantly reduced problems at startup. When we started up the South Korean joint venture, we depended on Portugal to do the initial preplanning for us, while we were in the process of training our engineers, and they did an excellent job. We experienced relatively few problems in starting up a new plant, with people totally inexperienced in wiring harness production. A lot of this can be attributed to good up-front planning and training, including the fine job that was done in preplanning all of the conveyor lines in Portugal.

After I left Packard and consulted for several years, I found out that most companies use a system for developing new manufacturing systems and production plans that had much more in common with the system used initially in Europe than the preplanning system that Packard had developed and used so successfully. I consulted with a large automotive component supplier that was relatively new to the wiring harness business, and they wanted to dramatically grow this business segment. Their main line of business was the manufacture of automotive seats.

To grow their wiring harness business (primarily by taking business away from competitors), they knew they needed to get more competitive. One way to do that was to reduce some of the onerous problems they encountered every time they started up a new product line. The managing director of their Mexican Operations was somewhat familiar with Packard's preplan process and was very open to setting up a similar system in their operations. I worked with them over a period of a few months to install a Methods Lab, train their engineers, and establish an enhanced preplan process, and they were very pleased with the results. They not only obtained significant reductions in labor content, they dramatically reduced the problems they had historically experienced on production startups.

They were so pleased with the results obtained from wiring harness preplanning that they wanted to initiate a similar program for their automotive seat business. Even though the products are very dissimilar, the principles used in the preplanning process are the same, even though the steps vary to some degree. We went through the same process that had been used for the wiring harness business, which included installing a Methods Lab, training their industrial engineers, and installing a preplan system fine-tuned for automotive seat manufacturing.

They used these tools to effectively preplan a new automotive seat program for a paced operation in one of their plants with great success, again demonstrating that the principles and concepts are universal and will help ensure that any manufacturing system developed, through utilization of this tool, will provide the optimum results with minimum problems.

But, Is It Worth It?

One would think that the following question would never be asked at Packard: Is the preplanning process worth it? Preplanning was responsible for hundreds of thousands of dollars of productivity improvements on each manufacturing system that went out into the plant, not to mention the tremendous savings attributed to reduced startup problems related to quality, delivery, efficiency, and the attitude of the workforce. And, the last item listed might be the most important in the list. Hourly workers expect salary workers to be as effective in what we produce as we expect them to be.

What message are we sending to hourly workers when there is the typical disaster surrounding the introduction of a new manufacturing system? The message they are receiving is that those highly paid college boys (and girls) don't have a clue about what they are doing. And, what do you think this does to their attitude about their jobs and about the company? It certainly doesn't motivate them to be better employees. I'm confident this phenomenon exists, and I don't know how to put a price tag on it, but I know it is a considerable factor in the cost profile of any company.

However, this question was asked, for one, by the "empty suit" who was largely responsible for the IPS I (Integrated Production System) and IPS II debacles (discussed in Chapter 3). It also was questioned by a few others through the 1970s, 1980s, and 1990s. Some speculated that we could use automated standard data to develop the production plans and to determine individual station breakdowns (similar to how it was being done in Europe, only with the introduction of automated standard data). The "suit" and others conjectured that any labor savings that had been achieved throughout the preplanning process could be gained on the plant floor through continuous improvement.

The problem is that their theories do not match the realities of what transpires on a production floor. Once a wiring harness conveyor system is installed, the flexibility to make significant change to that system is very limited—schedules must be met, and any significant change adds disruption to the manufacturing system, which cannot be tolerated. You either get the improvements up front by preplanning them into the system, or you lose most of the improvement opportunities forever. You can conduct all of the PICOS, Lean, or Six Sigma workshops you want, and you will never be able to recoup the savings that would be obtained from doing the job right to begin with instead of trying to "fix" it on the production floor.

Again, this is a prime example of some Packard executives being penny wise and pound foolish. The thinking was: "We can save tens of thousands of dollars in reduced industrial engineers' and methods operators' salaries if we eliminate the preplan process." However, that would mean annually throwing away millions of dollars in unrealized reduced labor content and reduced problems associated with production startups, which preplanning would provide.

There are certain products for which standard data (or one of the various Predetermined Time Systems) can be utilized effectively—products that require very discrete and simple movements to build. But, even in these cases, the entire manufacturing system, including the layout and flow, will not be established as well as could be done in a preplan process. For a wiring harness manufacturing system, the question is not even debatable. Even a very inexperienced engineer and operator team will produce a product that is significantly superior to anything that can be produced by even the most experienced engineer through the use of standard data.

Figure 7.1 is a graph that represents the benefits of preplanning. The x-axis is labor content for a product and the y-axis is years in production. When a product is preplanned, the initial labor content (when the product is first introduced to the plant) is very near the "Minimum Labor Content Line" for the product (unless the product has a significant redesign or some new technology is introduced). Over a period of time, the Preplanned Labor Content Line will approach the Minimum Labor Content Line through continuous improvement (although oftentimes, with a superior preplanning job, the Minimum Labor Content Line is achieved from day one).

However, for a product that has not been preplanned, the initial labor content is well above the Minimum Labor Content Line. Over time, the labor content will come down to some degree through continuous improvement,

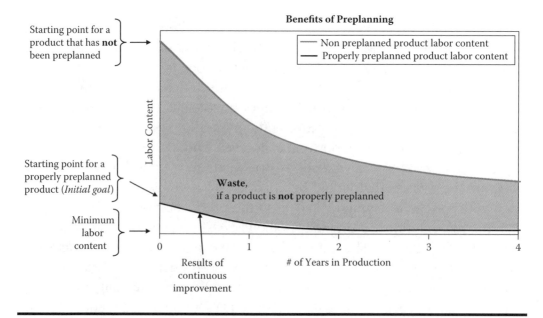

Figure 7.1 Labor Advantage Gained through Preplanning.

but the resulting Nonpreplanned Labor Content Line will never approach the Minimum Labor Content Line, in large part due to the lack of flexibility in the plant to implement significant changes to a manufacturing system in current production.

The difference between the Nonpreplanned Labor Content Line and the Preplanned Labor Content Line is waste, real waste, in dollars that a company is throwing away, and this waste is a whole lot more money than what a company would spend to accomplish proper preplanning on its products.

Packard Evolves, but Not for the Better

Due to a lack of understanding of the full benefits provided through the preplan process on the part of some Packard executives, there was a clear reduction in the emphasis and importance placed on preplanning throughout the 1980s and into the 1990s. There also was tremendous growth occurring in the Mexican Operations at this time, and it became very difficult to provide all of the industrial engineers with the fundamentals they needed, not only in preplanning, but in other aspects of the job as well. In addition, good IE supervisors were being transferred and promoted, on a regular basis, into other positions that needed to be filled, therefore, skills were even eroded further in this area.

All of these factors combined to substantially degrade the quality of the preplanning in the Mexican Operations (preplanning was almost extinct in Warren at this time, because most of the final assembly manufacturing had been moved to Mexico). Preplanning was still being done in Mexico, thank goodness, or things would have been much worse, but an upgrading of Mexico's preplanning capabilities was in order.

After I had returned from South Korea and was assigned to the Mexico East Operations, I quickly realized that much of the IE knowledge and many of the IE disciplines had been lost or were badly degraded. When I could find time, I started writing and accumulating an IE manual, utilizing much of the information I had developed overseas. A good chunk of the manual covered the entire preplanning system. One of the few weaknesses of Packard's preplan system from the early 1970s was in the lack of standardized forms. I knew that forcing the utilization of standardized forms would be very helpful in reinstituting the disciplines needed within the preplan system, so I established standard forms for each step of the preplan process. Eventually, I had all of these forms automated, at least the ones where calculations were necessary.

There was a resurgence and improved performance in preplanning in some areas, although it was not consistent throughout Mexico. What was needed was leadership at the top of the organization, ensuring that everyone was totally recommitted to reinstituting and revitalizing the preplan process, but it never happened. Preplanning continued to be done throughout the 1990s within Packard, and probably to this day, but never again at the level of excellence to which it was initially done.

In fact, the preplan process barely dodged a bullet in the year 2000. Delphi Automotive executives were obviously realizing that the company was in serious trouble and knew something had to be done to drive down costs quickly, or they were not going to be able to compete in the market. Delphi Automotive executives decided that the way to accomplish this was through the adoption of MSD (Manufacturing System Design) concepts. None of the other Delphi divisions had benchmarked Packard and did not have tools in place to accomplish what was accomplished through the preplan process at Packard.

MSD provided some tools to help accomplish Toyota's Rule #1 (all work will be highly specified as to content, sequencing, timing, and outcome). However, for establishing the optimum manufacturing system for a wiring harness, MSD, and the accompanying forms, was "a poor man's preplanning." The problem was that Packard executives didn't recognize this fact. Preplanning, as utilized by Packard, was much more specific, pertinent, and beneficial for developing the optimum manufacturing system for a wiring harness than was MSD, which was much more generic and did not include many of the necessary steps. But, since some Packard execs didn't understand this, a commitment was made to the bosses in Michigan that MSD would be implemented.

Ill-informed and politically concerned Packard executives were on the verge of creating another catastrophe. Fortunately, however, I solicited the aid of one of the few Packard execs who was knowledgeable in manufacturing and engineering, and who was not a "yes man." I explained to him that MSD, as was being proposed by Delphi and Packard executives, was inferior to the preplanning process, that it was much less sophisticated, specific, detailed, and pertinent for wiring harnesses than the system we had in place, and that to convert from Packard preplanning to MSD was going to create a disaster.

Fortunately, he agreed and he as able to convince the pertinent Packard executives to explain this to the appropriate Delphi executives. Thank goodness they were successful and the directive for Packard was rescinded.

But, why in the world should the effort to get the MSD decision cancelled have been necessary? It all goes back to lack of leadership and a number of managers who did not understand the business because they didn't have the on-floor experience, along with the politically motivated desire to accommodate their bosses.

At the end of this chapter is an outline of my refined preplan process for establishing optimum manufacturing systems for wiring harness production. Each of these steps requires the use of one or more forms (which are not included), to ensure that all of the steps in the process are successfully accomplished. The important things for a reader to understand are the principles involved, which are universally pertinent to any manufacturing operation. Steps can be modified, added, or deleted, and specific forms can be developed as needed, in order to accommodate any type of product or manufacturing system.

The Critical Four M Relationship

To have a truly effective manufacturing system, every element in Figure 7.2 must be in balance. If any of the elements are out of balance, improperly designed, or not performing as planned, the overall system will not be optimal, which can result in a company suffering various levels of inefficiency and inability to meet internal and customer expectations.

Effective preplanning is what will give a company a well-balanced and well-planned system. If preplanning is done properly, a company will have well-designed boards, tools, and equipment (*Machine*); components and

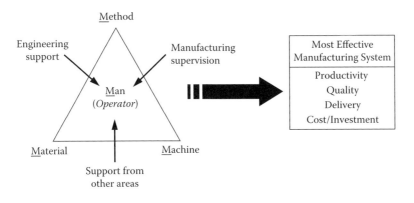

Figure 7.2 The Critical Four M Relationship.

circuits *cut to the proper length* in the proper locations (*Materials*); a proven and efficient *Method* that is ergonomically sound; and trained operators (*Man*) who can produce quality harnesses at the required rate.

Once the optimum manufacturing system is designed and installed by the industrial engineer, with the support of the other members of the Multidiscipline Team (MDT), it is ready to be released and started up in the plant. Effective organization and supervision by Manufacturing, combined with effective follow-up on the part of the industrial engineer and methods operator, as well as from the other members of the MDT, will ensure success in the plant.

Wiring Harness Preplanning

Following is a brief description of the steps involved in a preplanning process for developing a wiring harness manufacturing system, which should be helpful in gaining a basic understanding of the overall process. The fundamentals and principles of preplanning are universal for virtually any type of product and manufacturing process, although the specific steps and formats used will require customization for other products and businesses.

1. **Create Control Sheets:** Control Sheets ensure that important model change data and activities are properly documented and communicated. The quality and timeliness of these controls will give a good indication of the competency and the professionalism within the IE organization, and, if the communications are effective, the probability is good that the entire preplan process will go smoothly. For wiring harness manufacturing, a total of six Control Sheets were designed and utilized:
   – The Model Change Forecast, which provides an overview of key model change information for every harness package within each plant.
   – The Model Change Schedule, which communicates the scheduled build-out and startup dates for each package within a plant, along with other key model change information.
   – The Master Preplanning Schedule, which communicates all of the key events associated with each preplanning program within a plant.
   – The Final Assembly Process Equipment Requirements, which details all of the on-line equipment and tooling required in each Final Assembly Department as well as all of the offline equipment and tooling.

- The Conveyor Production Plan Information, which provides all pertinent data on conveyor part numbers, number of lines per package, number of crews per line, number of operators per crew, line speeds by part number, capacity per line, overall capacity percentage, etc., for all conveyors within the plant.
- The Stationary Board Production Plan Information, which provides standards by part numbers, capacities, etc., for all stationary boards within a plant.

2. **Develop the Labor Estimate:** The work content (*total direct labor content*) of the entire harness (*from cutting through packaging*) is estimated by the Cost Estimating Department using standard data. This estimated work content is used to develop costing and pricing, estimate investment and manpower requirements, and establish a Plant Operating Budget, thus, accuracy is very important. If the business is awarded (*won*), the labor estimate also is used in the development of the preplanning schedules. Because this information is so critical, and because industrial engineers have much more production floor experience than the typical cost estimator, when possible the preplan engineers (*or the lead methods engineer or the methods supervisor*) should review the estimates in some detail to ensure that no obvious errors or oversights exist.

3. **Develop Preplanning Schedule:** The lead methods engineer develops a preplan schedule for each harness program based on the estimated work content, the degree of change from the existing product, the number of part numbers, and the volume. It is important that these schedules are reasonably accurate and consistent to provide each industrial/methods engineer (*preplanner*) and methods operator the time necessary to do a professional preplanning job. (*To expedite this process, and to ensure accuracy and consistency, I designed an Excel®-based Preplan Timing Matrix, which develops a schedule for each preplanning program.*)

4. **Harness Design Analysis:** The Preplan Team (*the preplanner and the methods operator*) analyzes the harnesses within a package to determine improvements that can be made in harness design in an effort to reduce labor or materials, improve quality, and reduce or eliminate potential material handling or servicing problems. This could be a change in splice location, a change in the type of splice, changing from a splice to a double or vice versa, using conduit instead of tape or vice versa, using snap-on clips instead of tape-on clips, combining outlets, etc. (The customer may not agree to all of the proposals, but

if a proposal is not made, the answer will always be "no." Generally, if the customer can be convinced that the proposal will result in a better product or a reduced cost, the answer will be yes.)
- **DFM (Design for Manufacturability) Review:** When possible, a formal DFM review is held with the involvement of the Multidiscipline Team (MDT), with the intent of further improving the harness design. The MDT is comprised of the Preplan Team (the preplan engineer and the operator), the Lead Methods Engineer, the lab supervisor when available, and representatives from Quality Control, Product Engineering, Production Control, and the department foreman. Other participants could be added as appropriate, depending on the program and any special requirements.

5. **Develop a Pegboard and Organize a Pegboard Review Meeting:** A sample composite harness (*containing all of the features of all of the different part numbers within the package*) is placed on a pegboard in various configurations, by the Preplan Team in an effort to determine the optimum layout in which to construct the Master Board in order to accomplish the objectives of **minimizing** labor content, ergonomic concerns, interferences, potential quality concerns, board construction costs, and board size. The MDT reviews the pegboard proposal and suggests improvements.
   - **Design and Build the Master Board:** A Master Board is constructed per the final design determined in the Pegboard Review Meeting using the criteria established in order to minimize the board cost, while ensuring that the board construction will support good methods, good ergonomics, and quality. The preplanner should spend ample time with the board builder to ensure that the board is built as desired and that the appropriate board parts are utilized.
   - **DFM/DFA Review:** After the Master Board has been built, a formal DFM/DFA Review (Design for Manufacturability/Design for Assembly Review) is held with the MDT. During this review, a sample harness is placed on the Master Board and all potential build and/or quality problems are identified, regarding both the Master Board and the harness. Solutions are proposed by the group and corrective action is implemented.
   - **Order Repair Tools:** All new components are identified and a determination is made as to whether or not proper repair tools exist. If not, they are ordered from the appropriate supplier.

6. **Establish the Build Sequence:** An optimum build sequence (*which includes all wires, components, and harness coverings*) is determined by the preplanner, which should minimize harness interferences and plugging penalties, provide methods development flexibility, group compatible work elements, and reduce walk. For example, in large connectors, the ideal plugging sequence is from bottom to top (*to avoid visual obstruction*) and from left to right (*to avoid hand interference, because most operators are right handed*).
   - **Plan Pre- and Post-operations:** The preplanner makes a determination of all of the work elements or processes that will be done in the Final Assembly Department, but not done directly on the conveyor line or on the stationary board. Methods, layouts, material flow, and standards will be established for these processes later in the preplanning process.
   - **Establish the Circuit Test Sequence:** The preplanner establishes a ring-out sequence that ensures that all wires are electrically tested and that all necessary presence checks and special tests are made. For electro-mechanical boards (*meaning that the ring-out will be accomplished on the conveyor line as the harness is being built*), it is critical that the Circuit Test Sequence (CTS) match the build sequence.
7. **Establish Work Content:** The Preplan Team establishes the work content for each part number within the conveyor package (the total amount of time required by the methods operator to build a complete harness including time to inspect attributes, remove the harness from the board, and properly dispose of it to a packaging container or a subsequent operation). The Master Board is in a stationary position during this step, and while establishing the work content, it is critical that the methods operator is working at a 100% effectivity (speed and efficiency of movements), based on the British or Standard Scale as defined in Figure 7.3.

   This is one of the most critical steps in the entire process. It is virtually impossible to develop an effective production plan if an ineffective job is done in this step. Accurate work content becomes the basis for establishing a proper production plan—one that provides sufficient capacity to meet customer requirements, but one that does not contain excess capacity that promotes waste of resources.

   Much of the true methods analysis and determination will take place during this step. To determine the best method to accomplish

Scales					Description	Comparable Walking Speed (1)		Time to Deal 52 Cards (2)
60-80	75-100	100-133	0-120 REFA	0-100 Standard/ British		Miles/hr.	Ft/sec	Sec.
0	x	0	0	0	No Activity	0	0	0
40	x	67	60	50	Very slow, clumsy, fumbling movements; operator appears half-asleep with no interest in the job	2	2.93	52.0
60	75	100	90	75	Steady, deliberate, unhurried performance, as of a worker not on piecework but under proper supervision; looks slow, but time is not being intentionally wasted while under observation	3	4.40	34.7
80	100	133	120	100 (Standard Rating)	Brisk, business-like performance, as of an average qualified worker on piece-work; necessary standard of quality and accuracy achieved with confidence	4	5.87	26.0
100	125	167	150	125	Very fast; operator exhibits a high degree of assurance, dexterity, and coordination of movement, well above that of an average trained worker	5	7.33	20.8
120	150	200	180	150	Exceptionally fast; requires intense effort and concentration, and is unlikely to be kept up for long periods; a "virtuoso" performance only achieved by a few outstanding workers	6	8.80	17.3

Source: Adapted from a table issued by the Engineering and Allied Employers (*West of England*) Association, Department of Work-Study.
  (1) Assumes an operator of average height and physique, not carrying a load, and walking in a straight-line on a smooth, level surface without obstructions.
  (2) Assume a normal person with typical experience in dealing playing cards – not a Las Vegas card shark.

The British Standard Scale (0 – 100%) is the scale used by Delphi Packard and is the one which makes the most sense to me. The REFA Scale (0 – 120%) was used by Packard Reinshagen in Germany, but it was very confusing to use even for the supposed experts. We did an experiment in Portugal with several of the Portuguese engineers and some visiting engineers from Germany, and operators which were consistently rated 100% by me and the Portuguese engineers on the British Scale were rated anywhere between 100% to 150% by the German engineers on the REFA Scale.

Figure 7.3   Principle Operator Rating Scales.

each of the various work elements (*or logical combinations of elements*), it is important that various methods alternatives be investigated for each. Sufficient cycles are run on each element, for each of the viable methods alternatives, such that consistent, representative times are obtained. Additional Master Board and harness design improvement ideas also are frequently developed during this phase, which should be incorporated into the work content data (*assuming customer agreement is obtained for the proposed harness design changes*).

Work content is established in a similar manner for Stationary Boards, except that all of the different materials required to build the harness are serviced to the station. Also, the method, layout (*including how the materials are to be serviced to the workstation*), and standards are developed during this phase for all splices, subassemblies, and other pre- and post-operations for which there is no standard data, thus ensuring that these processes are considered an integral part of a well-planned manufacturing system.

8. **Develop the Production Plan and Conduct a Work Content Meeting:** Potential viable conveyor production plans are developed by the preplanner based on the customer requirements and total conveyor work content. The required number of operators, number of lines, number of crews, and line speeds are calculated from these two pieces of information. While the number of operators and the number of lines required by the system are somewhat fixed, several viable options generally exist regarding the number of crews that will be utilized (*a crew is defined as a group of operators who build a complete harness*) and the resulting line speeds. For example, a line can be set up with 16 operators with a single crew with a line speed of 25 seconds; it could be set up with two crews of 8 operators each with a line speed of 50 seconds (*each crew building a complete harness on their side of the line*); or it could be set up with 4 crews of 4 operators, each with a line speed of 100 seconds (*two crews on each side of the line, each building a complete harness*). Often the difference in investment for these various options is minimal and the right answer becomes a matter of preference based on the flexibility and practicality of the various options. (The various conveyor production plan alternatives are reviewed in the Work Content Meeting, along with information on all of the off-line processes. Because the entire manufacturing system will be determined in this meeting, it is the most important meeting conducted during the entire preplan process for each package. Therefore, participation by the entire MDT is important, especially from manufacturing supervision ([the foreman, etc.]). When they have a voice in the process and agree with the final decision, they will be much more committed to making it work on the floor.)

- **Distribute Data:** The Work Content Meeting form and/or Stationary Board Data form are completed and distributed by the preplanner at the start of the Work Content Meeting.

9. **Develop the Paper Breakdown:** Using the established build sequence and the work content times as a basis, the preplanner divides the work as evenly as possible among the stations, as established in the production plan (i.e., the number of operators per crew and the goal line speed), in an attempt to develop the best work element division and method to construct the harness. Sufficient time should be allowed in the last station to enable the operator to record defects and make minor repairs, as necessary.

10. **Run the Preliminary Breakdown and Conduct the Preliminary Breakdown Meeting:** Starting with Station 1, the preplanner and the methods operator work together to develop smooth-flowing stations, from the various elements of work content assigned to each station, as established in the Paper Breakdown. Once a viable method is established, the station is run on a preplan conveyor with the board moving clockwise at the goal line speed. A variety of methods should be evaluated for each station until the most efficient method is determined. When consistent (*level*) times are achieved (*again, while the methods operator is working at 100% effectivity*), the average station run time should be somewhere between the planned station time and 5% less than the planned station time (*a moving board, and the combination of some work elements within most stations, will generally create some improvement*). If the goal station time is not reached, it generally means that the method requires further refinement or that the methods operator needs more cycles to reach 100% effectivity. If the improvement is much more than 5%, it generally means that insufficient cycles or inefficient methods were run in Work Content. In extreme cases, it may be necessary to readjust the production plan so that excess capacity and investment is not built into the system. (After analyzing the results of all of the stations from the Preliminary Breakdown, the preplanner will make any necessary adjustments required to maintain an even division of work between stations (and to maintain a good flow within each station [this is called a "paper rebalance"]). A meeting is held with the MDT to review the results of the Preliminary Breakdown and to determine which stations need to be run in the Final Breakdown. The Preliminary Breakdown Meeting form is used as the basis for the meeting.)

11. **Run the Final Breakdown and Conduct the Final Breakdown Meeting:** The stations that have been adjusted in the paper rebalance, following the Preliminary Breakdown, are rerun in Final Breakdown to ensure an even distribution of work and to fine tune the final layout and method. After all required stations have been run, a meeting is held with the MDT to make a final review of the manufacturing system. The Final Breakdown Meeting form is used as the basis of the meeting. (Sometimes, due to late engineering changes or other factors, there is insufficient time to accomplish a complete Work Content, Preliminary Breakdown, and Final Breakdown. Of all of these steps, the Work Content is the most critical, and a complete running of Work Content should not be sacrificed, if at all possible. A good Preliminary Breakdown is the second most important step. If we have done a good job in the Preliminary Breakdown (*including making any necessary paper rebalances after all stations have been run*), we can be reasonably confident in the manufacturing system installed on the production floor, even if a complete Final Breakdown cannot be accomplished.)
    - **Distribute Data:** Following the completion of Final Breakdown, Conveyor Data and Packaging Data forms are completed and distributed.
    - **Finalize Visual Aids:** The preplanner and methods operator will develop all of the required visual aids for the boards, racks, conveyors, and pre- and post-operations.
12. **Develop the Department Layout:** A total department layout will be developed that includes all processes within the department, i.e., the conveyors and stationary boards, along with all of the off-line operations. The layout will be conducive to good material flow and material handling.
    - **Develop the Line Layout:** A detailed Conveyor/Stationary Board layout is developed that shows precise dimensions and how each wire and component is to be serviced.
    - **Issue Rack Order:** The preplanner fills out and issues a Rack Order form that will ensure that all required racking is available for model change installation.
13. **Finalize the PFMEA and Flow Diagram:** The PFMEA and Flow Diagram development will occur throughout the entire preplan process to ensure that quality risks have been identified and that measures have been taken to minimize/address these risks. (*This was added as part of QS9000 certification.* ***The preplan system worked*** *very **effectively prior to its introduction**.*)

14. **Accomplish the Master Board Release:** The preplanner, method operator, and quality assurance engineer jointly review and release the Master Board, using the established procedures, to ensure that it will produce a dimensionally, functionally, and electrically correct harness. A Master Board Release Sheet will be prepared and issued.
    - **Develop the Wire Length Letter:** This team will also determine wire lengths that properly fit the Master Board and allow for good harnesses to be built when the method is correctly followed. The wire lengths are determined as part of the release procedure.
    - **Build the Duplicate Boards:** Exact duplicates of the Master Board are built for regular production, as determined in the production plan. The Master Board is generally not used as a production board, but it may be used as a repair board on electromechanical lines.
15. **Provide Labor Standards:** The preplanner will provide a Direct Labor Bible, (DLB—see Chapter 6) (*or accurate standards for each process in the department*) to the foreman for each part number, as well as a complete Direct Operator Requirement Plan.
16. **Accomplish the Manufacturing System Release:** The entire manufacturing system, including the conveyor(s) and all related stationary boards and off-line operations, is installed per plan and then inspected to assure that everything is installed as designed and is functioning properly. Harnesses are built on some boards to ensure they will build a quality product. Assemblies also are made and inspected on each off-line operation and piece of equipment. The foreman and universal operators are trained during this process.
    - **Accomplish the Plant Installation:** The conveyor, duplicate boards, stationary boards, racks, and all tooling and equipment are installed in the plant in agreement with the conveyor and department layouts and the Model Change Schedule. A checklist form is used to record all open items and to assign responsibility. At-line meetings are held frequently to ensure that the installation is progressing as needed.
    - **Provide a Foreman's Model Change Notebook:** The preplanner provides the foreman's Notebook that contains the DLBs, visual aids, layouts, and other documents needed by the foreman to help him/her understand and manage the new manufacturing system.
17. **Accomplish the Line Start:** This is the start of training for regular production operators in the plant. The preplanner, methods operator, foreman, and universal operators train the conveyor operators station by station, as well as provide training on all off-line operations tied to

the line. Offline operators are trained a day or two prior to the start of the conveyor training in order to ensure that all necessary materials are available for the startup.

- **Hold a Line Start Meeting:** The preplanner will organize and coordinate meetings with all of the key individuals involved in the startup of offline operations and the line (*and/or stationary board*), including the direct operators, in order to communicate information about the program and to make the final preparations for the start of line training.
- **Accomplish Line Follow-Up:** The preplanner and the methods operator continue to work with the conveyor line after startup to resolve production and quality problems until the line is running efficiently at the line speed goal and producing good quality. Tools, such as Stop and Reject Studies, Cycle Checks, and Master Studies, will be used during the ramp-up period. Daily production and quality numbers should be recorded and graphed. Regular follow-up meetings should be held, with the preplanner and foremen reporting on progress made, plans in place, and support needed.

This process has been modified and enhanced quite a bit since the early days at Packard, but the principles are still the same. It was used very effectively to develop manufacturing systems that approached the minimum possible direct labor content and ensured smooth and effective production startups. The excellent work done in the Methods Lab almost functioned as a production pilot run because the simulation done in the lab was so near to actual production conditions.

As I indicated, the steps that I designed for preplanning the automotive seat conveyors, along with the formats, were somewhat different, based on the nature of the products, but the concepts and principles were the same. An effective manufacturing system for virtually any product can be established through the principles and concepts of the preplanning process, which can alleviate a lot of heartburn, not to mention costs, associated with the typical introduction of a new manufacturing system.

Chapter 8

The Computer Is a Moron

I realize that the title of this chapter is rather provocative, but I got the inspiration from none other than the renowned management guru, Peter Drucker. Drucker is reported to have said exactly this: "The computer is a moron." Some of you have undoubtedly felt this way as well (I know I have) when your computer seems to have a mind of its own and starts doing crazy things and popping up nutty messages all over the place. However, I don't think that is what Drucker was talking about. I believe that what he was talking about goes back go the saying: "Garbage in, garbage out."

I suspect that all of us, who have dealt with computers in any way in the business environment, can tell war stories about one or more programs that literally created more problems than they solved. How could the computer department (I'll call it Management Information Systems, or MIS, for short) be so boneheaded as to allow such an inefficient program to exist, and why don't they just fix it?

Based on my experience, the problem is that we put the blame in the wrong place. It's generally not the fault of MIS if the computer systems don't work as we would like. The problem is generally much closer to home. I will go back to a quote from Dr. W. Edwards Deming that pinpoints a good chunk of the problem. He said, "The problem is at the top; management is the problem." From what you have read so far in this book, you might be getting the impression, in the case of Packard at least, that this is the best guess response as to why any problem exists within a company.

It is management that makes the decision to purchase a gigantic, complex computer system, which has been designed and programmed by individuals not the least bit familiar with your particular business, which then expects the system to run smoothly in your operation and give you a competitive advantage. My question is this: How is any system that is bought off of the shelf, which can be purchased by anyone else, including your competitors, going to give you a competitive advantage? And, if your answer is that you are going to modify and fine-tune the system to meet your particular needs, did you really need to buy this big system to begin with? I remember when many companies years ago were spending a king's ransom to purchase SAP (Systems Applications and Products) software, which was so complex and not necessarily geared to their particular business that they could not get it to run properly. There were better alternatives.

It is management that makes the decision to keep an old and inefficient software system and then spend a fortune to try to upgrade it to the point that it will do what you need it to do. I thought I had seen some horrible computer systems until I accepted an assignment to Packard Hughes Interconnect and the Alabama Operations. After seeing what they had, I realized the other systems I was classifying as horrible, in fact, were only very bad. This one was horrible. It was 1960s technology trying to service a year 2000 business, and it just did not have the flexibility and capacity to do it. Management made the unbelievably bad decision to spend $2 million dollars to upgrade this system, primarily due to a very persuasive computer engineer at the company headquarters in Irving, California, who was the company's expert in this system. Alabama engineers knew this was the wrong answer and fought hard against it without success, that is, until $1 million had already been futilely spent; then it was finally acknowledged that, even if successful, all we were going to have in the end was a very expensive 1960s technology system that still wouldn't do what we needed. Finally, the correct decision was made to purchase a new system, at a fraction of the cost, which had the capacity and the flexibility to allow us to meet our specific needs. Yes, there was a better alternative.

It is management that makes the decision to tweak an ineffective program within a system (leaving a program that is only a little less ineffective) rather than spend a few extra dollars to totally redo the program so that it becomes an effective tool that will have a positive impact on the company's competitive position. I tried very hard in the late 1970s to get approval for MIS to reprogram our out-of-date and ineffective efficiency control system. I had redesigned it down to dotting every "i" and crossing every "t", and I am confident that a programmer could have redone the entire program in a few weeks, but, no dice. What we got was a one-day "lick and a promise" that wasn't really worth the effort. I am confident that if the program had been rewritten, thereby greatly improving its effectiveness, there is no way the entire productivity control system would have been tossed out in the mid-1980s, a decision that was, of all the misguided decisions made by Packard management, the undisputed champ. Yes, there was a better alternative.

It is management that makes the decision to pull the plug on a new program, which they have agreed could save tens of millions of dollars per year because of a budget crunch, thereby saving a few thousand dollars in programming costs. When Packard management decided it would not proceed with the reintroduction of an efficiency (productivity) control system, totally redesigned and updated with the latest technology, after agreeing that this type of system could help the company save at least $50 million yearly, it became the last straw for me. If a company is not willing to spend a couple hundred thousand dollars once in order to save at least $50 million annually, that company's future is not that bright, especially when upper management's rationalization is that the company just couldn't afford it. I know that Packard was wasting a lot more money than that on projects that were not going to save a nickel. For example, the company spent $7 million on tooling and equipment for a one-piece flow process in the truck plant that actually decreased plant productivity by half (detailed in Chapter 4). Yes, there was a better alternative.

However, it is not just management that is to blame for the failures of many computer programs and systems, it is also the fault of the users (and sometimes these are one and the same). My experience is that most users have a totally unrealistic view of what MIS's responsibilities should be to ensure the success of computer programs versus what their own responsibilities should be. Many users expect that the MIS personnel should not only be experts in computer systems, but that they should be experts in all other aspects of the business as well. Some employees also must assume that they are mind readers, based on some of the written requests I have seen made of the MIS Department.

One thing is for sure, I am never going to be a great programmer. Sure, I made an A in Fortran 40 years ago in college, and I can do a little simple programming on Excel®, but that is about the extent of it. I would never be a great programmer regardless of what training I received; it is unrealistic to think otherwise. The same goes for most great programmers. They are never going to understand all of the aspects of the business as well as the people who do these jobs on a daily basis.

One of the things I have discovered over the years is that everyone's mind works differently. What seems simple and straightforward to one person can be very difficult and confusing to another, and the tables can be reversed when considering other issues. Programmers tend to go into programming because it interests them and their minds work in such a way that programming is fun and easy for them (or, at least it should be). The rest of us have gone into other fields for the same reasons. Of course, it's a big plus if an individual knows and understands things outside of his area of expertise, but don't ever expect that a programmer is going to know as much about your area of expertise and your specific computer needs as you do.

A Rare Opportunity

When I was asked to go to South Korea and assume total operational responsibility for the startup of a new joint venture plant that was manufacturing wiring harnesses for the local and export market, I had a very rare opportunity and I took full advantage of it. Over the years, I had been very critical of computerized systems and programs in use at Packard in the various operations with which I had had experience. Most of them didn't tell me what I really wanted and needed to know to do my job more effectively, and many of the reports contained a lot of information that was of no use.

In this new position, I had an opportunity to change all of that. We spent a couple hundred thousand dollars on a basic HP2000 system with standard software, but we did not have the specific software programs and systems that we would need to run the operation effectively. We hired an MIS supervisor, who had very good programming skills, and he confirmed that the system we had purchased provided all of the capacity and flexibility we would needed to develop and program the computerized systems that we desired.

Based on my experience (and, hopefully, wisdom), I knew that, in order to run an effective operation, we would need systems specific to our needs for automated cutting, efficiency (productivity) control, estimating and pricing, material control, quality control, personnel control, crib control, and financial control (plant operating reports, P&Ls for part numbers and plants, etc.), as well as a host of other small programs that could make life easier, such as die control, suggestion control, fiscal inventory, etc.

I knew that there was one way, and one way only, to obtain these systems and programs, and that was to design them myself with great specificity and then work hand-in-hand with MIS to get them programmed. The programs we needed just did not exist on anyone's shelf, nor could we expect that they would. Each of these systems and programs would need to be very specific to our requirements, and our requirements were just not the same as anyone else's, at least to some degree. Packard certainly didn't have the programs we needed or wanted (although they should have wanted the same ones I wanted).

Some of you are probably thinking that it was quite presumptuous of me to think that I was going to be able to just sit down and design all of these systems. Most of my past experience had been in engineering and manufacturing, but the systems I mention above went well beyond those two disciplines, and I had no programming experience. The good news is that you don't need to know how to program in order to design a system, you only need to know how a computer works. If you understand that a computer is a repository of information that must be input and, that once input, this information can be manipulated and then spit out in the format you specify, you know everything you need to know about the computer. As far as not having direct experience in areas outside of engineering and manufacturing, while this was true, these two functional areas worked closely with the other functional areas mentioned, as well as the fact that just plain common sense will dictate what is important and needs to be accomplished within the other areas, so long as you have a good basic understanding of the business.

At the start of the process, I wrote an introduction to a book I intended to maintain, which would contain all of the systems and programs that I would need to design and develop, and on which I would need to work hand-in-hand with MIS to make a reality. I wanted to make sure my MIS philosophy was clear and I wanted to ensure that everyone in the MIS Department knew my expectations and understandings. I wrote this back in the summer of 1986, but after reading it again, there is not much I would change.

> "The computer is one of man's greatest inventions and if it is used properly it can be a fantastic tool to help organize and manage a business and drive down costs. Unfortunately, if the computer is not used properly, the opposite results can and do occur.

The computer is an expensive tool to buy, maintain, and service. Computer technicians and programmers don't come cheap. The only way, therefore, that a computer can be justified is:

1. To reduce headcount in other areas serviced by the computer sufficiently to offset the MIS headcount and pay off the investment and reoccurring costs of the system.
2. To create savings in other cost areas through the use of the computer, i.e., improved productivity, reduced premium shipment costs, reduced inventories, reduced turnover and absenteeism, improved quality, etc.
3. To perform essential jobs that cannot be done (or done in a timely or quality manner) manually.
4. Or, to accomplish a combination of the above.

In most businesses of any size, a computer system can be easily justified using the above criteria. Unfortunately, in many cases where computer systems are justified and installed, the expected benefits never reach the rosy projections, and sometimes the benefits are, in total, negative.

Why does this occur? Besides the fact that all of us tend to be overly optimistic at times, especially computer people concerning the benefits of a computer system, I believe the biggest one reason is due to the totally unrealistic expectations on the part of users concerning the development of computer programs. Many users expect the MIS department to not only program a system for their area—a task for which they are very qualified—they also expect them to design the system, an assignment for which they are generally very unqualified.

It is simply not reasonable to expect that the computer experts in a company also are going to be experts in every other area of the company for which systems need to be designed. The only people who really know what information and processing of information are required are those people who work in the various areas—the users. While the MIS Department will certainly provide design assistance and advice, the majority of the system design detail must be developed by the user; this will include formats for the final report(s), database information required, and input information required. If this is left up to the MIS

Department, with only general guidelines being given concerning what is wanted, the results will be very disappointing and the computer may turn out to be a burden rather than an asset.

Below are some guidelines concerning computer systems that I believe can help ensure that the computer helps us to compete rather than prevents us from being competitive:

1. Programs and systems must be designed by the users (to include all final reports and all input screens), with assistance from the MIS Department.
2. Insist, as a user, that the program does exactly what you want it to do and that the reports appear exactly as you wish them to. Don't be talked out of something you want or need just because you are told that it is not easy, or even that it is not possible. Be convinced that it cannot be done and then look for other angles to get at the same results.
3. With rare exception, it is better to start from scratch and design your own program(s) for your specific needs, assuming the resources exist to do so, than to try to fit your needs to an existing program or try to make big modifications to an existing program.
4. Don't create a computer program if it's not justified, even if the computer system is in place. Each program should stand on its own. Unnecessary programs will cause a loss of credibility for all programs, even the good ones.
5. When designing a program, try to put the maximum amount of information on the minimum number of reports. The more reports generated in a program or system, the higher the cost and the less likely they are to be used effectively. Also, try to minimize the number of different input screens required and use the same input in as many programs within a system as possible."

Based on my experiences in South Korea, where we had some outstanding programs that helped us become a more efficient and cost-effective operation versus some of the other programs with which I have had experience, I believe this is still valid and good advice.

The last thing I want to do now is to get into a lot of detail about the many computerized programs (generally being comprised of a single input screen and a single report, with limited requirements to access the company database) and systems (generally comprised of more than one input screen and more than one report, with significant requirements to access the company database), so I won't. We have already reviewed the efficiency control system that we put in place in South Korea in Chapter 6, and I will review the cutting system we put in place in Chapter 10. However, I would like to review a couple of programs that we put into place that were extremely important in helping us manage an effective operation, and in the process demonstrate two things: (1) you do not have to be a computer expert to design a relatively large and complex computer system, as long as you know what a computer is able to do and if you have a good understanding of the processes and activities for which you intend to develop the system; and (2) you do not have to be the ultimate authority for the processes and activities involved, so long as you have a good basic understanding and use some common sense.

Controlling Absenteeism, Turnover, and Overtime Is a Must

After spending a year and a half in Industrial Engineering (IE), during which I preplanned several rear body harnesses and engine harnesses for two model changes (my first assignments at Packard), I was asked to take an assignment as a foreman in one of the Packard Warren Engine Harness Departments. I was excited about the move, because I knew that if an individual wanted to really understand the fundamentals of a manufacturing business, including how to manage people and overcome problems, the best way to do it was to be a first-line supervisor in manufacturing. I also knew that my IE experience would be very helpful, especially with my thorough understanding of the efficiency control system, because great emphasis was placed on efficiency in those days at Packard and rightly so, especially with the high cost of labor.

Unlike my experiences when hiring into IE, wherein I received a very thorough and detailed training, I got the standard one-day "here's the department, good luck" orientation that I think most everyone else received. The foreman, whom I was replacing, spent a day with me explaining how things worked or, at least, how things were supposed to work. And, at the end of the day, told me he was going to his new assignment the next day, with the promise that he would be close by and would be more than happy to help if I called (I didn't).

I also got a bit of an orientation from my new boss, the general foreman, but the plant could certainly have taken a lesson from IE as to how to train new employees. As I have come to learn more and more about Toyota, I'm pretty sure they have a little different strategy on how to train new foremen and ensure they operate with the correct principles and strategies. Packard did use one concept to train new foremen who were promoted from the hourly ranks, which was very helpful, but not complete. Most of the training involved assigning the new prospect to work with an experienced foreman for one week, and expecting the prospect to learn through osmosis. While this experience was necessary and good as far as it went, this concept should have been the frosting on the cake, not the cake itself.

One of the things I realized very early was the very negative impact that absenteeism and tardiness had on the smooth and efficient running of the department. The only information a foreman received on absenteeism was a monthly absenteeism report, with no detail as to what encompassed these numbers. The absenteeism rate for my department had historically been about 10%, which was about average for the plant as a whole. But, just having this one number was not of much help in trying to reduce the problem to a manageable level. I knew I needed detail, on an operator by operator basis, if I was going to quickly make a dent in the problem.

I tried to get this detail, but I was told that it was not available. I knew that it was unavailable only because no one would authorize MIS to develop the report. All of the needed information was in the database. Even back in 1973 it would not have been that difficult to program a simple report for each supervisor with month-to-date (MTD) and year-to-date (YTD) absenteeism for each hourly operator. Cutting the absenteeism rate by half (which I knew this type of report could help accomplish) would have resulted in millions of dollars of annual savings in improved productivity and reduced headcount, but this did not motivate management to authorize its creation, either because they didn't understand the potential benefits or they were too engaged elsewhere.

My only recourse was to develop my own report. I designed a spreadsheet using legal-size graph paper on which I kept a daily record of the absenteeism and tardiness performance of each operator. An amazing thing happened once I started keeping the report—absenteeism started falling like a rock (not so amazing once one understands that "when you measure performance, performance improves"). When I was making the chart at my desk, a couple of universal operators saw what I was doing and asked me about it. When I told them, the word quickly spread throughout the department. That month, the department absenteeism dropped to around 6% and then to 5% and, ultimately, to 4%. All I had to do was talk to the one operator who had the worst attendance performance, and that single act motivated that operator, as well as every other operator in the department, to improve his performance.

But, it sure would have been nice to have a computer report, especially because a history would have been available as well. This would have freed up some of my time to do more productive things within the department.

When we started up the South Korean joint venture, I was determined to give production supervision at all levels, myself included, the tools needed to properly control absenteeism and turnover. In the MIS manual, I explained the importance of these controls:

> "Maintaining proper control of hourly workers, especially direct hourly employees, is extremely important if an operation is going to be productive and cost effective. Three of the most important factors to control are absenteeism, overtime hours worked, and turnover rate. High levels of absenteeism (in total, or for any specific individual), sustained high levels of overtime, and high turnover invariably lead to:
>
> 1. Excess headcount resulting in increased burden
> 2. Lower efficiency levels due to:
>   a. Constant training to replace departed operators
>   b. Substitution of less qualified and/or experienced operators for missing operators
>   c. Decreased morale and increased operator fatigue
> 3. Lower quality levels due to:
>   a. Use of new and/or inexperienced operators
>   b. Use of experienced operators on unfamiliar jobs
>   c. Decreased morale and increased operator fatigue"

The process of creating these computer tools is quite straightforward and not overly complicated. The first thing I did was to take a spreadsheet and design the reports that I wanted, exactly as I wanted them to appear. The report for individual operator performance was designed to go to the specific department foreman, although higher levels of management could ask for the report if desired. Summary reports were designed to be distributed to higher levels of management.

Once a report was designed, I determined how each piece of information would be determined or calculated by the computer and printed on the report. Some of the information would come directly from data already in the database; some would come from calculations made from information already in the database; and some would come from information that would have to be input or from calculations made from information that would have to be input, in conjunction with database information. The bottom line is that, if you want information on a report, you can have it. You just have to figure out where the information is going to come from and make plans to have the necessary information available, either from what is already in the database or from input.

Figure 8.1 is a re-creation of the report I designed for the control of individual operator's performance, with a few actual examples from the South Korean plants. Summary reports were actually done on separate reports, but for simplicity, I added an actual summary at the bottom of the operator report. To help understand the report, Total Absenteeism includes all absenteeism for any reason (excluding scheduled vacation). Net Absenteeism backs out excused absenteeism for military service (required of all male South Koreans), excused days for marriage, and excused bereavement for the death of a family member (all nonpaid).

Date: 12/31/87				Attendance, Turnover, and Overtime Report								Foreman: K.S. Park						Supervisor: D.S. Lim	
Dept: Quality Assurance				Month-to-Date Data								Year-to-Date Data							
Oper. #	Name	Level	Sex	Military Service (*)	Days Absent	Times Late	Total Hours Scheduled	Total Hours Worked	Absenteeism%	Excused Days	Net Absenteeism%	Overtime Hours Worked	Overtime%	Days Absent	Times Late	Total Hours Scheduled	Total Hours Worked	Absenteeism%	Excused Days	Net Absenteeism%	Overtime Hours Worked	Overtime%
1252	K.S. Lee	2	M	*	4	0	248.5	222.0	10.66%	0	10.66%	48.5	21.85%	24	14	2822.5	2629.0	6.86%	0	6.86%	524.5	19.95%
1045	B.S. Park	2	M		2	0	244.5	227.0	7.16%	0	7.16%	52.5	23.13%	11	6	2786.5	2679.0	3.86%	5	2.14%	556.5	20.77%
1017	Y.W. Kang	1	M		6	0	274.0	222.0	18.98%	6	0.00%	58.0	26.13%	20	7	2861.0	2861.0	6.29%	15	2.10%	589.0	21.97%
5106	Y.S. Kim	1	F		0	0	282.5	286.5	0.00%	0	0.00%	70.5	24.96%	9	3	3033.0	2951.0	2.69%	3	1.90%	686.0	23.24%
1012	J.H. Lim	1	M		1	1	260.0	247.0	5.00%	0	5.00%	52.0	21.05%	9	3	2866.0	2794.5	2.49%	5	1.10%	554.0	19.82%
5127	M.H. Cho	2	F		0	0	282.0	286.0	0.00%	0	0.00%	70.0	24.82%	2	4	2801.0	2776.5	0.87%	0	0.87%	481.0	17.32%
5080	S.H. Kim	1	F		0	0	275.5	275.5	0.00%	0	0.00%	59.5	21.92%	0	0	2843.0	2843.0	0.00%	0	0.00%	496.0	17.47%
etc.																						
Resignations														6	2	2799.5	2750.0	1.77%	0	1.77%	449.5	16.35%
5025	M.S. Jung	1	F		2	0	268.0	246.0	8.21%	0	8.21%	52.0	21.14%									
1046	C.S. Joun	1	M		0	0	9.5	9.5	0.00%	0	0.00%	1.5	15.79%									
Department Total					15	5	2919.0	2756.0	5.58%	6	**3.94%**	583.0	21.15%	82	57	28823.5	28039.0	2.72%	34	**1.78%**	5508.5	19.65%
Annual Turnover					199.92%									23.04%								

Summary from 8/31/90																		
YangJu Plant **Direct** Hourly	460	306	64471.7	60574.5	60.4%	317	1.88%	9513.5	15.71%	2475	2154	422860.3	401080.5	5.15%	1008	3.05%	67233.0	16.76%
Annual Turnover	10.92%									40.08%								
YangJu Plant **Indirect** Hourly	38	32	7152.8	6847.0	4.28%	29	0.98%	1353.0	19.76%	327	251	50555.7	47979.0	5.10%	196	2.04%	10112.5	20.87%
Annual Turnover	0.00%									9.72%								
Total JangJu Plant Hourly	498	338	71624.5	67421.5	5.87%	346	1.79%	10895.0	16.16%	2815	2411	473416.0	449059.5	5.14%	1204	2.95%	77635.5	17.29%
Annual Turnover	9.36%									36.12%								

1. This data is from actual reports that I kept as examples. This data for the quality Assurance Department and the summary data for the YangJu plant were from different dates. On the actual Summary Report, data was available for both plants, as well as the overall data for the company.
2. The data for any department could be sequenced in any order desired. This data is sequenced from highest Net Absenteeism YTD to lowest.
3. Data for all persons leaving the company during the year are shown under YTD. Those leaving during the month are shown under MTD (the two resignations during December were two of the three, or possibly four, resignations for the year for this department. One turnover rates compared very favorably with typical South Korean turnover rates of over 60%.
4. Overtime typically ran between 15% and 20% (generally 9.5 hour day and a full day on two or three Saturdays per month, versus the half day which was part of the normal South Korean schedule). If excessive overtime had have had an undesirable impact on absenteeism or productivity, additional operators would have been hired; but as it was, we determined that it was more advantageous to have fewer workers with somewhat higher levels of overtime than it was to hire additional workers.

Figure 8.1 Attendance, Turnover, and Overtime Report.

In addition to this information, which is very important in helping to run an effective operation, I knew we could use the information already in the computer to provide additional insight. We could use the computer to identify performance by categories (in whatever categories we wanted to consider, which we felt might have an impact on performance). The cost of absenteeism is very high, but the cost of turnover also is very significant. I have seen studies that conclude that it costs several hundred dollars to replace a Mexican operator who has resigned, and the cost would be higher for a South Korean operator, and much higher for operators in higher wage areas. The turnover rate was historically very high in South Korea for hourly workers (about 60% per year; not as high as the 100%+ per year in the Mexican Operations, but very high, none the less).

Analyzing data on absenteeism and turnover by category in South Korea (as well as in Mexico) makes a lot of sense when you realize that this data can be used in the hiring process to help select individuals who, based on historical data, will have a greater chance of success. South Korea and Mexico do not have EEOC (Equal Employment Opportunity Commission)-type guidelines in place that would prohibit a company from making an objective analysis of turnover and absenteeism data by various categories and then using that data as a part of the selection and hiring process. In South Korea, the categories we considered were sex, marital status, military status (for men only), and dorm status (women only). What we found is that women (both married and nonmarried, regardless of whether they lived in the dorm or not) had about half of the rates of both turnover and absenteeism that men had, as well as the fact that they were much less prone to create havoc. We used these data to tilt our hiring toward women, who were more consistent in the workplace and who were less likely to quit or miss work.

When I returned to the United States, I tried to get our personnel director to see the merits of using the computer to help determine the ideal hiring profiles, which could dramatically reduce our turnover rate that was then in excess of 100%. Because we had tens of thousands of hourly workers in Mexico at the time (something over 30,000), if it only cost $300 to replace each of the 30,000 workers during the year, the company was spending around $9 million dollars annually related to turnover. Even a reduction of 10% could mean real money, and I believed with better selection criteria, it could be reduced a lot more than that. The categories I suggested be considered were sex, age, marital status, parenthood status, education level, living arrangements, distance from work, transportation to and from work, prior work experience, and home town, with maybe one or two other factors being thrown in that were deemed to have an influence on predictable performance. Some of this information was already available on the job application form and the rest could be added. The computer could then be used to establish actual performance by the various categories considered. This information then could be used to aid in the selection process and potentially dramatically improve results.

I didn't get anywhere with the personnel director, so I submitted a suggestion through the Suggestion Program. As you might guess, this was a total waste of time. My experience with Packard's Suggestion Program is that suggestion dollars were paid out to individuals who made suggestions to make their jobs more convenient, but that saved the company no money. Any suggestions that had the potential to save the company a lot of money, but that would require a little energy to evaluate and then to implement, never made the grade.

The South Korea operation is just one case where the computer was used to create reports that, in fact, did save the company a lot of money by reducing absenteeism and turnover. I designed and we implemented all of the other systems mentioned earlier in this chapter, and each of them provided a very valuable service that did help us be a more effective organization. I will only touch on one of them briefly, because it was a critical program and one definitely outside of my areas of expertise.

Material Availability Is Critical

I never had the opportunity to work directly in Material Control, but having worked most of my career in Industrial Engineering and in Manufacturing, I had the opportunity to work frequently with the Material Control Organization. I also heard MRP (Material Requirements Planning) used frequently when in discussions about needed materials and what might be expected. What was quite apparent is that MRP was supposed to ensure that the plants received the materials they needed, when they were needed, and in the quantities needed. Packard seemed to have a lot of difficulties over the years in making this happen. I never quite knew if it was the system or the man behind the curtain that was responsible for the problems—a combination of both, I assume.

In any case, we did not have access to MRP or to any other computerized materials program in South Korea. What we did have was a German expatriot who was on a short-term assignment to Shin Sung Packard with responsibility for the Material Control Department. He was a very hard working and knowledgeable materials guy, who had spent his entire career in Packard's European Operations in the materials area, which was very helpful because about a third of our materials were coming from our German operations. It was quite impressive watching him work on his manual spreadsheets, just like they must have had to do things in the days before computers. Because our operation was still fairly small, he could establish requirements with pencil and paper, but I knew this was not a long-term solution. He would soon be gone, and we were scheduled to get much bigger, and we were already experiencing problems receiving the materials we needed in a timely manner.

I spent a little time talking to this engineer to make sure my understanding of the requirements of an MRP system were correct, and then I retired to my office to design a system that would work for our specific needs (to include all reports and input screens that the system would need). It actually came together a lot easier than I thought it might, and I have no idea how it compared to what Packard was doing (not very close, I suspect), but, it did what we needed based on our particular circumstances, which were very much different than Packard's. After doing a lot of thinking, I designed the system in a couple of days, and then I reviewed it with the German engineer who blessed it and said that he couldn't wait to get started using it.

We then assigned one of our two programmers to the project. The first order of business was to educate him as much as possible in a couple of days as to what materials planning was and what we needed to accomplish (the MIS supervisor was also part of this training). We reviewed each of the reports that would be required, each of the input screens that would provide needed data, and the information that was already in the database, which would need to be accessed. Once he had a basic understanding, he started the programming, with the understanding that I would work very closely with him during the entire process. As it turned out, that entire process took about two months. Six or seven weeks of that time was programming, which was basically done by one technician, and the other couple of weeks were used for debugging the system. In the end, we had a program that was designed specifically for our needs, and we were able to add some features that I doubt are common to most systems (maybe to no other system).

It cost us a few thousand dollars to create this system, tailor made to our specifications. I wonder what we would have spent if we had purchased a materials management system and then had to modify it as needed, or if we had contracted an outside company to program the system for us using our specs. Based on later experience, I'm sure the cost would have been a large multiplier, and we would still not have had the in-house expertise needed to properly maintain and update the system as needed over time.

The basics of any materials planning system are the same. A forecast is made of future material requirements based on stock list information, which is a detail of all raw materials needed in the manufacturing process by part number (in our case, this was already in the database as it was needed for costing and pricing purposes) and forecasted build requirements (in our case, based on the weekly shipping requirements given to us by our customer, Daewoo Motors). These forecasted requirements then are used within the company to plan and manage raw materials based on company policies and practices.

We were getting some materials locally in South Korea, some from Japan, and much of our materials came from Packard's European and Warren Operations. In all cases, we decided to use surface freight to keep the costs of transportation down. We would use air fright for emergencies only (which fortunately never came, in large part due to the MRP system we installed). Surface delivery time, including all customs and clearing time, was eight weeks from both the United States and Germany, two weeks from Japan, and a couple of days from our South Korean vendors. In-transit materials became one of the three "buckets" of materials that we tracked within the system, and the amount of material within this bucket was determined by the shipping location of the supplier.

The second "bucket" of material was what I called in-process material. This also was dictated by the supplier based on the length of time it took from the reccipt of the order until that material physically left the supplier's door. This period of time varied from supplier to supplier, but in most cases it was about four weeks. This only left the third bucket of material, which was the inventory of material we maintained in our raw material warehouse. In a perfect world, this could have been one week (or even a few days), but this isn't a perfect world. Forecasted schedules change, especially when you are forecasting out 12 weeks and beyond just to cover the time necessary for order processing and shipping. Also, suppliers don't always ship what they are supposed to or when they are supposed to.

Taking everything into consideration, we established a plan to maintain four weeks of material in the warehouse coming from the United States and Germany, three weeks of material coming from Japan, and two weeks of material coming from within South Korea. We knew that it would be easy to adjust these targets if experience dictated it was prudent. The one thing we knew for sure was that there would be no forgiveness if we missed a shipment to the assembly plant attributable to missing raw materials. And, we also knew that one or two premium air freights from the United States or Germany would more than offset an entire year's savings attributed to one fewer week's inventory in the warehouse.

What our system allowed us to do was to track every piece of raw material through the entire process—from when the order was initially placed until that material arrived at our plant warehouse door. For example, assume part number A is coming from Warren and further assume that this week's forecasted volumes from the customer did not change from the prior week. Four weeks' requirements of this material would be in the warehouse at the beginning of the week, eight weeks of this material would be in-transit, four weeks of this material would be in-process, and we would be preparing to order the material required for week 17 (rounded to the next standard pack, through an automated material order form generated by the system).

However, forecasts changed frequently, so what the system did was to compare, for each material part number, the quantity of material in each of the three buckets versus what the new forecasts indicated should have been. The computer then took these three variances into account in determining the new order quantity (i.e., week 17 requirements plus or minus the sum of the three variances). If the forecasts had decreased, the actual order might be zero (possibly for several weeks if the decreased requirements were significant). If the forecasts were increased, the week 17 requirements would be increased by the positive variance, and then the system would check to see that sufficient material was in the pipeline to ensure that no material outages would occur prior to receiving the additional materials. If there was a possible material availability problem, the system would flag the respective material part number so that special attention could be given to develop a solution.

If results are any indication, this home-grown system was extremely effective (not to mention very cheap). In the almost four years this system was up and running in South Korea while I was there, we never once had to shut a line down for missing materials and our premium shipment costs were zero. Of course, this system also gave us the value of all warehouse inventories. I believe we could have started to draw down our warehouse inventory levels by plan, based on our experience and success, and thereby saved a few bucks, but the cost avoidance we obtained by eliminating premium freight and costs associated with jeopardizing customer delivery would dwarf these savings.

We did another interesting thing with this system, which is something you can do when you design your own system. We established a Supplier Rating System, which gave a numerical score to each supplier (monthly and a year-to-date) based on that supplier's adherence to meeting shipping dates and quantities requested, and another score based on the quality of materials received in the warehouse. This information was sent to each supplier, along with a scatter chart showing the performance of all other suppliers (without naming them). We saw a steady improvement in the performance for our suppliers, which is directly attributable, I believe, to the fact that we were measuring supplier performance and then reporting on that performance.

Summary

Remember these universal concepts (truths), they are valid everywhere:

- What you do not measure, you do not control.
- When you measure performance, performance improves.
- When you measure and report performance, performance improvement accelerates.

The computer is a wonderful invention and, if it is used wisely, it can be a great tool to help drive improvement. But, a computer must be used wisely and that doesn't mean going out and buying the most expensive hardware and software you can find. My contention is that most companies, large enough to have an MIS Department, would benefit much more from specialized programs and systems designed for their specific needs by knowledgeable and innovative users, and programmed by their own MIS personnel, than by trying to find something on the shelf that will "do the job."

Once your own employees program a system, they become the experts who will then be able to maintain and enhance the system efficiently and cost effectively. If you contract an outside firm to program a system, it will be very expensive and the ongoing maintenance and updating costs will be exorbitant, not to mention less convenient, timely, and secure.

Chapter 9

How to Drive Down Total Process Cycle Time (TPCT) without Wasting a Lot of Time (and Money)

Virtually everyone who has had any introduction to Lean Manufacturing concepts has undoubtedly been introduced to the concept of Value Stream Analysis. This was a tool used quite intently by Packard starting in the mid-to-late 1990s, with the express purpose of driving down the Total Process Cycle Time for both manufacturing and nonmanufacturing processes. Before I provide information on the level of success achieved by Packard through this process, let me describe it briefly for those unfamiliar with this concept.

The first thing to understand is that, whenever there is a product or service provided for a customer, there is a value stream (I also like to call it a *critical path*). The challenge lies in understanding and improving it. Value Stream Mapping is a tool that allows an individual to see and understand the flow of material and information as a product or service is processed through the various departments within a company (i.e., the value stream). The total time taken by this process is called the Total Process Cycle Time (TPCT).

For manufactured products, a value stream begins with the receipt of raw materials and ends with cash in hand as the customer pays for the product. For nonmanufactured products, the value stream begins with the identification of a customer's or team member's need and ends with the delivery of the information or service.

The first step in Value Stream Mapping is to draw the Current State by mapping each step in the process using a set of symbols or icons to represent all processes, inspections, moves, flows, and stores (inventories). These maps can become quite complicated as there are often several different branches flowing into a single step in the process. A cycle time is established for each step in the process and a TPCT is determined by adding up the individual cycle times for the longest branch.

The second step in the process is to analyze the Current State and determine where improvements can be made and waste eliminated. The Future Value Stream should then be mapped.

The final step in the process is to prepare an implementation plan that describes how the Future Value Stream will be accomplished. As the plan is implemented and as the Future State becomes reality, the mapping process repeats itself as there is always opportunity for continuous improvement (especially if you don't do it right the first time).

Some of you can probably already guess how this process worked for Packard. If you guessed "not very well" you win. As with many things Packard did related to Lean, the focus was put on creating and "using" the tool, not on the benefits that could be achieved through its use. However, in Packard's defense, the tool, in most cases, is not actually all that helpful.

The first problem is that the actual processing of material is only a small percentage of the TPCT (sometimes as low as a percent or two). Most of a TPCT is made up of inventory time, so the solution to reducing TPCT was always to reduce inventory, whether or not that made economic sense. Secondly, Packard was taught that even though dozens of branches might be feeding one step in the process (e.g., dozens of different components and materials feeding one manufacturing step), only one or two of the more critical (more expensive) ones needed to be considered. This made it possible for Packard to focus on only one or two items, at the exclusion of all of the others. When I suggested that it might make a lot more sense to consider an average inventory for all of the materials feeding the step or even the total dollars of inventory feeding the step, I was told that that was not how it was supposed to be done (in other words, do it like the book says, whether or not it makes any sense).

I'm not aware that anything positive ever came from Packard's use of Value Stream Mapping other than giving engineers practice on using their computers for developing some neat charts. In reality, the types of processes on which Value Stream Mapping was used within Packard were not all that critical in the total scheme of things, but TPCT was a very important concept for some of the Value Streams that existed within Packard, and a tool to help drive these TPCTs down was urgently needed. Value Stream Mapping just was not the tool to provide that help.

Critical Value Streams

In contrast to the Value Stream as defined and explained above, I would like to define a Critical Value Stream as a sequential series of dependent actions taken by various departments or functional areas within an organization, required to provide a critical service or product to a customer. Packard had many of these value streams (or critical paths), but what it lacked was a good system to reduce the TPCT, which could have given Packard a competitive edge and really would have allowed Packard to "exceed its customers' expectations." This was a mantra heard many times, but rarely delivered.

For example, imagine any critical product or service required by a customer. The value stream or critical path usually starts with some type of order or notification from the customer. Let's assume that Department A is responsible for receiving the customer contact. Once Department A personnel receive this customer contact, they complete their task, then they provide critical materials or information to Department B personnel who complete their task; they, in turn, provide critical information or materials to Department C personnel who complete their task, and so on, until the last step in the value stream has been reached. It should be understood that there may be a lot of other activities going on in support of the value stream, but they are not necessary in order to define or even to measure the TPCT. The objective of the organization is to reduce the TPCT for this process to the lowest reasonable level, which means it is done within the values and principles of the company, and it is performance that can be repeated and maintained without extraordinary effort or cost.

Depending on the product or service and what the objectives are, a starting and ending point for the value stream must be determined. For example, if a company is trying to drive down the time required to obtain a new terminal die, the starting point might be the receipt of a blueprint from the customer indicating the need for a new die, and the ending point might be the delivery of a debugged die to the proper manufacturing department.

If a company is trying to drive down the time required to implement engineering changes, the starting point might be the receipt of a marked print from the customer detailing the requested change, and the ending point might be the date that the customer receives material containing the new changes in his or her warehouse.

If a company is trying to drive down the time required to provide prototype production to the customer, the starting point might be the official order received from the customer, and the ending point might be the arrival date of the prototype production at the customer's location.

For each of these examples, a different starting or ending point could be chosen, depending on the objectives to be accomplished and the current problems being encountered. For example, the value stream could start with an informal document from the customer that could precede the official document normally used. Or, if a company is having difficulty receiving compensation for services or products rendered, the value stream might end with receipt of payment from the customer.

Packard "Attacks" the TPCT for Engineering Change Implementation

Over the years, Packard used the "task force" concept many times to try to solve various problems. Some of these task forces were quite successful, but many were not, largely dependent on the makeup of the teams and the nature of the tasks given. One such task force was organized in the mid-1990s to take on the very important task of driving down the time it was taking Packard to implement engineering changes.

In the old days, Packard's average time to implement engineering changes was 8 to 10 weeks, which wasn't great, but it was a lot better than the 14 weeks it was taking in the mid-1990s. The new general manager wanted to see the average implementation time driven down to two weeks, although no one seemed to know where this number came from, or even, necessarily, what the starting and ending points for the process were to be.

The Engineering Change Task Force was headquartered in Warren, but, because of my expertise in Methods and Industrial Engineering, I was asked to provide some outside assistance to the group. Meetings were held monthly for six months as I recall, and I got to sit in on a couple of meetings when I was in Warren on other business. The meetings were both interesting and contentious at times.

One of the problems was that because final assembly was now a minor part of the Warren business, a lot of the controls and disciplines that had once existed in Warren in the management of engineering changes had been lost or were certainly less robust than they had once been.

This was the case in the Mexico Operations, as well. For this reason, when I compiled the Industrial Engineering (IE) manual, I rewrote the Engineering Change Procedures, complete with all necessary forms, which I had automated using Excel.

Remote cutting and lead prep also were big obstacles to quick implementation of engineering changes. The other problem was that, depending on which end point was used, there were at least 7 different departments responsible for up to 22 sequential, dependent steps in the overall process, and each step was assigned an allowed accomplishment time period without regard to the type of change that was requested, i.e., there was no distraction in time allotted for a simple change vs. a very complicated change.

What generally transpired in the meetings was that each step in the critical value stream was reviewed, starting with step 1 and proceeding through the last step. For each step in the process, recommendations were given as to how the allotted time for that step could be reduced. (These recommendations were almost always made by team members who were *not* in the department responsible for accomplishing the step.) The person in the responsible department would then present a case as to why that would not work or why the time could not be reduced. The end result was that virtually nothing changed. When the task force started its work, the TPCT of the Engineering Change Value Stream was about 50 days, and, when it ended its work, it was 45 days—still a very long way from the two-week goal.

I am sure the participants in this task force could not have been all that proud of what was accomplished, but at least they could counter that they didn't commit themselves or their comrades to meet dates that might be difficult to achieve for certain types of changes. The problem is that once a time period is established to complete an activity, that is the length of time it will almost always take to do it, even if the circumstances are such that it should take much less time for certain types of changes. It reminds me of another saying: "Work will expand to fill the amount of time available to do it."

The pattern I saw played out by this task force is very similar to what I observed in several other situations, and I knew there had to be a better way to achieve the necessary goal of reducing TPCT for critical value streams.

Key Activity Control Is Born

Shortly after this task force was disbanded, with their less than stellar results in hand, I was assigned as the Business Segment Leader for the Mexican Operations for the F-car (the Firebird and Camaro, which were being built in St. Therese, Canada). One of the problems St. Therese had with Packard was the excessive amount of time it was taking to get engineering changes implemented and the new products to its door.

I knew that one of my first orders of business was to help get engineering changes under control and to provide assistance to help drive down the implementation time. Based on prior experience, I knew that the following three concepts could be used creatively to do just that:

1. What you do not measure you do not control.
2. When you measure performance, performance improves.
3. When you measure and report performance, performance improvement accelerates.

I knew without question that the TPCT could be driven down dramatically if these concepts were incorporated within the value stream analysis.

Before I detail what I implemented, let's review the process Packard used (and probably the process most companies use) to try to drive down TPCT for critical value streams. Packard's typical method can be outlined as follows:

1. Establish a task force or call a multidiscipline team meeting.
2. Review the current critical path and standard timing (TPCT) versus the objective.
3. Brainstorm ways to eliminate or consolidate steps and/or reduce the time required to complete each step.
4. Reach agreement on various initiatives.
5. Make assignments and agree on follow-up meeting dates.
6. Review results at the next meeting.
7. Results achieved?
   a. Yes: Celebrate (rarely, if ever, happened)
   b. No: Go back to step 2 and start over (or eventually just give up when the objectives cannot be met)

My observations are that what was generally accomplished through this type of system was:

1. Bruised feelings (you are playing in my sand box, i.e., you have got lots of great ideas on how I can reduce time for my activity, but none for how you can reduce time for yours).
2. Lack of ownership (it's not my fault we didn't meet the goal, I only own a small part of the process).
3. Lack of accountability (when everyone is accountable, no one is accountable).
4. Very slow process (no one feels any real sense of urgency the way the meetings were spaced out, and everyone rationalized that they had lots of other important things to do).
5. Historically, minimal improvement is obtained (like getting 5 days out of a 50-day value stream over a 6-month period, and this was only on paper; who knows if that paper improvement will ever be realized in real life).

I didn't care much for the way things were done, but if you are to complain, you better have a better idea. Fortunately, I did I have a better idea and I was able to use it several times to make meaningful improvements in TPCT. I will contend that anyone can use this tool in any type of an organization, manufacturing or otherwise, to dramatically drive down the TPCT for any value stream for any type of product or service. Each of the steps in the process is important, but step 6 is the step that separates it from what most others do, and this is the step that is responsible for most of the improvement.

1. Select an important value stream:
   a. One that has caused the most difficulty in allowing the company to meet customer's expectations.
   b. One that offers the most opportunity for improvement in responsiveness and/or cost reduction.
2. Establish a multidiscipline team (MDT):
   a. Include at least one open-minded and knowledgeable individual from each of the functional areas or departments in the value stream.
   b. Include one or more outsiders to participate, who should be completely open minded because they have "no dog in the fight."
3. Identify all activities in the critical path (value stream):
   a. These are only the dependent activities that must be completed, in sequence, for the value stream to be accomplished (these are the activities that dictate the TPCT).

b. **A succeeding activity can often be STARTED before the preceding activity is completed;** this is often possible and it is motivated by this system, which helps to greatly reduce TPCT.
c. Activities, other than the dependent activities, may be ongoing in support of accomplishing the value steam, but, if they are not in the critical path, they do not impact/dictate the TPCT.
4. Assign responsibility for each activity (or step in the critical path/value stream):
   a. Depending on the activity and the nature of the process, the responsibility can be assigned to a department, a discipline, or to a specific individual.
   b. Generally, the more specific the assignment of responsibilities, the better the ultimate results will be from this process.
5. Establish targets for improvement:
   a. The overall target is often dictated by customer expectations or by upper management dictates.
   b. When the answer has not already been dictated, an overall improvement goal should be established (i.e., a percentage reduction or a numerical reduction in the TPCT) based on a reasonable expectation of what can be accomplished, such that the effort and cost required to implement the process is justified.
   c. Once the overall improvement goal is defined, the number of goal days to accomplish the entire value stream should be equitably divided among the various activities in the critical path.
   d. This system will work even if agreement cannot be reached at this point on the amount of time allotted to each step (or each area or individual) in the critical path.
6. Develop a Key Activity Control and TPCT Performance Report to organize and control the process and to record and measure performance (in most cases, both can be captured on the same format).
   a. This is the key ingredient missing from the traditional approach—the tool that is responsible for the dramatic improvements in performance that we realized through its use. (Specific examples of this type of control will be given later in this chapter.)
   b. The bottom line is that with the use of this control, the performance for each individual or department on each of the activities for each of the programs or categories is detailed.
   c. Once the theory is understood, it is relatively easy to structure one of these controls (and to computerize it, if desired, to save time and

energy; most can be done on an Excel® spreadsheet). While the Key Activity Control will look different for every distinct value stream, the concepts are exactly the same. Once you have done one, you can structure a Key Activity Control for any value stream.

7. Review the performance on a program by program basis and identify opportunity areas:
   a. As the various steps in the process are completed for each program, the completion dates are to be entered into the system.
   b. Once all of the steps in the process have been completed for one or more programs, a TPCT Performance Report can be automatically generated (if computerized).
   c. The multidiscipline team (MDT) should monitor the performance on a regular basis and identify opportunity areas, i.e., activities that are taking the greatest amount of time and/or are taking the highest percentage of time over the goal time.
8. Focus on opportunity areas and implement continuous improvement:
   a. Because we are measuring performance by functional area or by individual, it will be apparent where improvement needs to be made and who needs to take responsibility for getting it done.
   b. While MDT discussions can be helpful, most of the improvement ideas will be generated by the responsible areas, because they should be the most knowledgeable about their specific area and functions.
   c. Because we are objectively measuring and reporting the performance by area or individual, and not just the end results of the entire process, there is tremendous **incentive on everyone's part** to help drive down the TPCT.
   d. To provide additional incentive, the performance reports can be distributed to others within the organization that have a vested interest in obtaining a TPCT reduction (e.g., upper management).
9. Follow up until all objectives are accomplished:
   a. This process will drive the TPCT to the lowest maintainable level, because each individual/area is striving to accomplish his/her step(s) in the process as quickly as possible.
   b. If targets cannot be met, the MDT needs to analyze why and try to remove or modify the roadblocks.
   c. If the goal proves to be unattainable, the realistic results need to be communicated to the organization so that unrealistic commitments are not made that will destroy credibility with customers, suppliers, or co-workers.

Once I developed this new concept, I quickly designed the necessary formats for engineering change implementation for the F-car Team. Until that time, the performance of the F-car Team was pretty much in line with the overall Packard performance, i.e., it was taking about 13 or 14 weeks on average to get wiring harnesses to the assembly plant that contained the new changes from the time the customer issued an official request. For a few very difficult changes, this might have been acceptable, but for an average change it was ridiculous.

You have probably heard the saying that when you mix apples and oranges you get a fruit salad. That's exactly what happens when you mix all of the various types of engineering changes into one big pot and look at the average implementation times; it doesn't tell you very much. One plant could have a better average performance than another plant, but that could be because they have a higher concentration of simple changes and not because they are more effective in engineering change implementation.

And, where in the world do you start trying to attack problems if all you have is one overall average TPCT for all engineering changes that have been implemented? The answer is that you can't do it unless you have much more detailed data. Remember: "Without data you are just another person with an opinion," and having one overall average number is not much better than no data at all.

I wanted to place engineering changes into as few categories as possible, while ensuring that sufficient categories existed such that changes could be properly grouped. Based on the steps that needed to be accomplished for change implementation and on reasonable expectations of what could be achieved regarding timing, I came up with four main categories, the last category having three subcategories.

1. Coordinated Changes: These changes needed to be handled separately from the others because **Packard did not control the timing of these changes**. The timing was controlled by the customer and was dictated by some other change being made in the vehicle to which we had to coordinate our change implementation. Certainly the timing of these changes should not impact the scorecard for the changes for which we did have control.
2. Changes Requiring New Materials: This means materials that were not currently in use at Packard. It might be a newly designed component that Packard had to design and produce, or it might be some component that had to be found and purchased from the outside.

In any case, there was significant time involved in obtaining the new materials, which would distort the TPCT if these changes were lumped with the other types of changes of which we had much more control over the outcome.
3. Changes Affecting Leads: This meant that, with remote cutting and lead prep, the Warren plant had to cut new leads and then send them to Mexico to be used in coordination with existing customer authorized lead inventories. (If Mexico had had full service plants, as we once did, this process could have been shortened by several days with little risk for scrapped leads. With remote cutting and lead prep, it became a much more time-consuming ordeal and the risk for scrap was much greater.)
4. Changes to Be Implemented Quickly:
   a. Normal Changes Not in Categories 2 or 3: These were normally simple changes affecting the external covering of the harness and could generally be processed through the entire value stream in a matter of a few days, as opposed to the several weeks it was typically taking.
   b. Urgent Changes: This meant that the changes were to be implemented as soon as possible and that the customer would pay for any scrap leads generated, so long as approved inventory levels were not exceeded.
   c. No-Build Changes: Meaning that the change was to be implemented as soon as possible and that all harnesses in the system had to be reworked.

I think you can see that lumping all of these changes together into one big pot, and then looking at an average implementation time, was not going to be very helpful in driving down the TPCT. Some of these changes only required a few of the 22 steps in the value stream, while others required all 22. Lumping all of the different engineering change types together provided a distorted view of reality.

I don't want to get into a detailed description of the Key Activity Control and the automatically generated TPCT Performance Chart for this system, because I will give a couple of other examples later in this chapter. What I will tell you is that we categorized each engineering change we received and tracked the time it took to complete each of the activities in the value stream for each change (from customer authorization date to the date when harnesses incorporating the new changes arrived at the customer's dock). This was not a cumbersome process, because the dates were already being reviewed in the Engineering Change Coordination Meeting, and the Key Activity Control was the document used to actually plan the changes in the meeting.

The end results were that we achieved an overall TPCT of less than eight days for all changes we processed for the F-car for 1996. For category 4 changes, the average TPCT was six days, understanding that almost half of this time was in transportation to the St. Therese assembly plant. For category 3 changes, the average TPCT was a little over two weeks. Remembering that we started the process with an average TPCT of about 13 weeks, which was normal Packard performance, this is a fairly significant improvement (almost 88%) (that is, if you consider 88% significant).

I presented this information, along with the system that had driven the improvements, at the next Business Segment Leader Conference. These conferences were held a couple of times a year and were attended by most of Packard's top operations executives. Everyone seemed to be impressed with the results as well as the process that had generated them. I received a lot of accolades at the meeting and I expected that I would be asked to help implement the system throughout Packard as a best practice.

However, the request never came. Shortly thereafter, I was asked to accept responsibility to manage our wholly owned subsidiary in Lower Alabama, which was in real trouble with its Boeing wiring business. That left no one in Mexico to champion this tool that had been responsible for such dramatic improvements. I later found out that even the F-car team stopped using the system, and their performance gradually returned to the Packard norm.

Over the years, Packard had paid a small fortune to various outside consultants for tools that were far less helpful than the one I had given them for nothing, but Packard management did not have the insight or common sense to use it.

Others Benefit from the System

After my tour of duty at Packard Hughes Interconnect in southern Alabama, I returned to the Mexico West Operations and was given engineering responsibility for Packard International, which was comprised of several plants that were providing harnesses to non-GM customers. One of the things I found out very early in my assignment was that we had a very crucial customer who was very unhappy with us because we were not able to meet his timetable of 15 days to provide prototype harnesses to his door once he had placed an official order. This customer was so unhappy that he was threatening to place his business elsewhere.

I found out that for several months, regular meetings had been held with a multidiscipline team to try to figure out how to drive the TPCT down to 15 days from the 40+ it was then taking. I sat in on the next scheduled meeting, and it was "deja vu all over again," as Yogi Berra would say. Everyone in the meeting kept trying to justify why they needed the time they were currently taking to accomplish their tasks, but almost everyone had a plethora of suggestions about what others could do to reduce their process time. The problem was that no one actually knew how much time it was taking to complete any of the steps in the process, because no control sheet or tracking device was being used. Everyone just knew that the whole process was taking too long, and each participant knew he had to defend his area and shove blame elsewhere. Does that sound familiar?

After listening to an hour or so of this, it became obvious that the group needed a lot of help and wasn't going to get there on its own. I told them that it seemed this team had a serious challenge in trying to meet the expectations of a very valuable customer. I said, "I have experience with a system that I am very confident can help get us to where we need to be. If you will work with me, I can commit to you with a high degree of confidence that we can drive the TPCT down to the 15 days expected by our customer in a relatively short period of time."

This team had been chastised enough over the previous months that team members were more than willing to give it a try. All I really needed from them was help in defining the critical value stream. Once we had done that, I told them that I would design a tool to help us achieve our goal, and I asked that we meet again the next day to review it and get started.

That evening, in an hour or so, I designed a combined Key Activity Control/TPCT Performance Report and rolled it out the next day at the meeting. One of the industrial engineers, who had been working on the prototype harness program, was assigned to maintain the Control and it was agreed that the IE supervisor would function as the team leader.

This format would be used to track every prototype program for this customer through the value stream (critical path), from the time a formal order was issued by the customer (which included a blueprint) until we had delivered the prototype harnesses safely to our customer's dock. Based on some of the discussions in the meeting and my prior experience on how long various activities took, I established goals for each of the steps in the critical path (knowing that agreement would never be reached in an MDT meeting), with the overall TPCT being 15 days, which was the answer dictated by our customer. I knew there would have to be some dramatic changes made in how some of the areas approached their responsibilities in the value stream, but I was confident the goal could be achieved with the proper motivation, which this system provides.

So, what happened? During the first three months this system was in place (January through March, 1999), the average delivery time had decreased from over 40 days (the 1998 average TPCT) to 20 days, but over half of the average TPCT (11.2 days) was being spent on the single activity of Material Control providing new materials to the plant after the legal prints were made available by Product Engineering. In the review meeting, after the first month's results were available, questions were asked as to why this activity took so long and the team brainstormed things that could be done to drive the time down.

It was revealed by someone in the group, possibly the product engineer, that this customer's policy was to provide advanced notice of any new materials required for proposed prototype builds several weeks before official orders were actually made (even if it was not their policy, we could have requested this information). This early material information was not 100% accurate, but it was probably 95% or more accurate. Because we were only buying small amounts of any new materials for the prototype builds, there was very little exposure even if we ordered a material that would eventually be scrapped. By ordering materials from the advanced notice, rather than waiting for the official order, several days (or even weeks) could be saved in the completion of this activity in the critical value stream. The resulting improvement in customer satisfaction, and the securing of this customer's business for the future, would certainly be worth a lot more than the minor risk and cost associated with a small amount of potential scrap material (seeing as how we had contracts worth millions of dollars annually that would be in jeopardy if we didn't get our act together on the prototype builds).

This potential improvement existed since the beginning of the pilot program with this customer, but it was never explored or brought to light. Why? Because there was no mechanism in place to shine a light on just how big a problem we actually had with this activity in the value stream (critical path). Once sufficient light was cast on this very big problem, a solution quickly followed.

On the May report (there wasn't a lot of consistency on cut-off dates that determined on which month's report a program would be recorded, but, no matter, they all got reported on one of them) the TPCT had been reduced to 13 days (we finally exceeded our customer's expectation), and the year to date (YTD) average had been reduced to 18.5 days (Figure 9.1).

Again, it is very interesting what happens when you start keeping score. Material Control had reduced their activity time to 5.6 days, half of what it had been the first three months the system was in place and a small fraction of what it had been the prior year.

IE and Manufacturing were completing the prototype build for each program in an average of less than one day. Impossible, you say? Yes, it would have been if they had waited until all of the materials were in the plant to get started. However, they realized that much, if not most, of the manufacturing could be accomplished even though one or more materials might still be missing. They could then complete the manufacturing process when the final material(s) arrived in a very short period of time. **It is just human nature that when you are being scored, you will find creative ways to achieve the best performance possible.**

The same scenario existed for Quality Control regarding the final inspection and packing of the prototype harnesses. If QC had waited until all of the harnesses were completed before starting their activities, as they had done in the past, they would have experienced little if any improvement. What they started doing was preassembling all of the necessary packaging materials and working hand-in-hand with IE and Manufacturing such that the final inspection was started at the earliest possible moment. In cases where harnesses were built except for missing material, inspection was completed except for the incomplete part of the harnesses. When the material arrived and the harnesses were completed, the final inspection was quickly completed. By taking these steps, they drove their cycle time down to the lowest realistic level.

For some of the activities, like delivering the harnesses to the warehouse or delivering the harnesses from the warehouse to the customer's dock, there was not a lot of opportunity to make improvements to the raw time it took to complete the process. What this system ensured was that harnesses were not sitting around somewhere, adding unnecessary time to the TPCT, because no one was paying attention to them.

By the July report, the overall TPCT for the month had been reduced to 7.9 days, and the overall average TPCT for the year had been reduced to 14.8 days. Product Engineering was turning around the customer prints

Company X Prototype Key Activity Control

Month / Year May-99

Model Year / Program	# of Part Numbers	# of Pieces Ordered	Avg. # of Circuits per Harness	Date Prints Received by Product Eng.	Product Engineering Provides Prints to MDT		Material Control Provides Materials to Plant for Prototype Build		I.E. and Mfg. Complete the Prototype Build		Q.C. Completes the Inspection and Packing of Prototypes		Material Control Delivers Prototype Harnesses to Warehouse		Traffic Delivers Prototype Harnesses to Customer's Dock		MDT Prints Received > Harnesses Arrive at Customer	Remarks
				Goal →	2 days		2 days		6 days		1 day		1 day		3 days		15 days	
					Date Compl.	Days	Date Compl.	Days	Date Compl.	Days	Date Compl.	Days	Date Compl.	Days	Date Compl.	Days		
99 - 41	22	32	174	03/17/99	3/18	1	4/8	15	4/9	1	4/12	1	4/13	1	4/15	2	21	
99 - 42	3	90	4	03/18/99	3/18	0	4/1	10	4/2	1	4/5	1	4/6	1	4/9	3	16	
99 - 43	3	6	252	03/22/99	4/6	11	4/6	0	4/6	0	4/6	0	4/7	1	4/12	2	15	
99 - 44	19	42	87	03/29/99	4/1	3	4/20	13	4/20	0	4/21	1	4/22	1	4/27	3	21	
99 - 45	1	30	2	03/30/99	3/31	1	3/31	0	3/31	0	3/31	0	4/1	1	4/5	2	4	
99 - 46	1	200	4	04/08/99	4/8	0	4/9	1	4/14	3	4/15	1	4/16	1	4/21	3	9	
99 - 47	7	18	28	04/09/99	4/19	6	4/29	8	4/29	0	4/30	1	5/3	1	5/5	2	18	
99 - 48	9	13	70	04/20/99	4/23	3	4/26	1	4/27	1	4/28	1	4/29	1	5/3	2	9	
99 - 49	1	1	339	04/21/99	4/21	0	4/30	7	4/30	0	5/3	1	5/4	1	5/6	2	11	
99 - 50	1	1	80	04/21/99	4/22	1	4/23	1	4/23	0	4/23	0	4/26	1	4/29	3	6	
MTD Avg.	6.7	43.3	104.0		MTD Avg.	2.6		5.6		0.6		0.7		1.0		2.4	13.0	Total MTD Avg.
YTD Avg.	5.9	45.0	101.0		YTD Avg.	3.4		8.2		1.9		1.0		1.1		2.9	18.5	Total YTD Avg.
MTD Total	67	433																
YTD Total	295	2250																

Remarks

Figure 9.1 Key Activity Control/TPCT Performance Report for Company X, May 1999.

the same day they received them. Material Control was now providing materials to the plant in less than two days, and the other areas continued to demonstrate strong performance in accomplishing their steps in the critical path (Figure 9.2).

The bottom line is that, through the use of this relatively simple tool, we were able to reduce the TPCT of producing and delivering prototypes to a very valuable customer from over 40 days to under 8 days (an improvement of over 80%) and, in the process, potentially saved the business, and not just the prototype business, but all of the business from this customer (Figure 9.3).

And, why did this happen? It is not, after all, an overly complex tool. It all goes back to two concepts: (1) when you measure performance, performance improves: and (2) you will get the performance you motivate. I can assure you, it is a lot easier and more productive to create this kind of a tool/process and let it motivate the members of the MDT to use their creativity to improve their performance and that of the total team, than it is to try to sit in a conference room, with everyone trying to protect their territory, and hash out a plan that will actually accomplish anything. (I also should note that an additional concept can be used to expedite even more improvement, which we did on this project, namely: When you measure and *report* performance, performance improvement accelerates.)

One would think that with the success we had achieved in driving down the TPCT for both engineering changes and prototype delivery (in excess of 80% in both cases), Packard would be salivating over this system and would want to implement it as a best practice throughout the operations, but, you would think wrong. It just didn't happen, even though the story was written up in the company paper as an example of a tremendous success in reducing TPCT. How could this happen? I can't say, unless one considers the possibility of incompetent management.

On the bright side, Packard upper management did agree to fund the programming of a fairly sizeable system (that utilized these concepts) that I designed to help manufacturing obtain new components in a timely manner for prototype builds, pilot builds, and for start of regular production. This had been an ongoing problem, and I knew we could solve it by measuring performance and holding people accountable. That's the good news. The bad news is that there was another budget crunch in late 2000, and the funding for this project, which was modest in relation to the advantages that could be realized, was cut and the project bit the dust. This was just another in a long line of short-sighted decisions made by Packard upper management.

Company X Prototype Key Activity Control

Month / Year July-1999

Program	# of Part Numbers	# of Pieces Ordered	Avg. # of Circuits per Harness	Date Prints Received by Product Eng. Goal →	Product Engineering Provides Prints to MDT 2 days		Material Control Provides Materials to Plant for Prototype Build 2 days		I.E. and Mfg. Complete the Prototype Build 6 days		Q.C. Completes the Inspection and Packing of Prototypes 1 day		Material Control Delivers Prototype Harnesses to Warehouse 1 day		Traffic Delivers Prototype Harnesses to Customer's Dock 3 days		MDT Prints Received > Harnesses Arrive at Customer 15 days	Remarks
					Date Compl.	Days	Date Compl.	Days	Date Compl.	Days	Date Compl.	Days	Date Compl.	Days	Date Compl.	Days		
99 - 75	5	5	266	06/22/09	6/22	0	6/24	2	7/2	6	7/5	1	7/6	1	7/9	3	13	
99 - 76	1	180	6	07/01/09	7/1	0	7/2	1	7/8	4	7/9	1	7/12	1	7/14	2	9	
99 - 77	10	17	140	07/01/09	7/1	0	7/7	4	7/8	1	7/9	0	7/12	1	7/15	3	10	
99 - 78	1	30	2	07/02/09	7/2	0	7/7	3	7/12	3	7/12	0	7/13	1	7/15	2	9	
99 - 79	1	1	1	07/07/09	7/7	0	7/7	0	7/8	1	7/8	0	7/8	0	7/12	2	3	
99 - 80	1	1	50	07/22/09	7/22	0	7/23	1	7/23	0	7/26	1	7/27	1	7/29	2	5	
99 - 81	3	38	3	07/26/09	7/26	0	7/27	1	7/28	1	7/29	1	7/30	1	8/3	2	6	
MTD Avg.	3.1	38.9	61.1	MTD Avg.		0.0		1.7		2.3		0.7		0.9		2.3	7.9	Total MTD Avg.
YTD Avg.	5.6	43.2	98.0	YTD Avg.		2.1		6.3		2.1		0.9		1.0		2.4	14.6	Total YTD Avg.
MTD Total	22	272		Remarks														
YTD Total	454	3499																

Figure 9.2 Key Activity Control/TPCT Performance Report for Company X, July 1999.

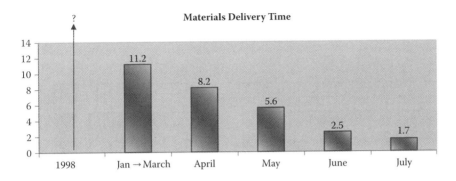

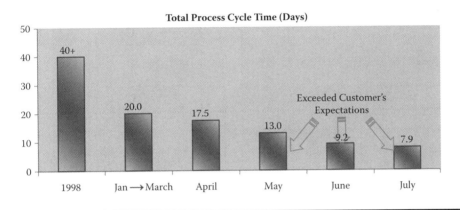

Figure 9.3 Results from using Key Activity Control/TPCT Performance Report System for Company X from 1998 through July 1999.

This System/Concept Will Work Anywhere

When I left Packard and started consulting with another large company, which was producing components for the auto industry, I found that it had the same problem Packard had regarding excessive TPCTs for various value streams. One of the big problems the company was having was in obtaining new terminal dies in a timely manner in order to meet manufacturing requirements. I designed a Key Activity Control to help resolve this problem. The system had only been in place for a few weeks when I finished my consulting assignment, so most of the dies had not yet been procured, but already it was evident that the time periods required to complete each of the various activities in the critical path, by the respective individuals or departments, were starting to show significant improvement (Figure 9.4). The one die that had been procured had been completed in less than the time set for the goal, which had been established by Process Engineering. A second report (Figure 9.5) was designed

256 ◾ *Intelligent Manufacturing: Reviving U.S. Manufacturing*

Figure 9.4 Key Activity Control/TPCT Performance Report for New Die Acquisitions for Company Y.

Supplier Performance on Die Delivery

Date ____ / ____ / ____

Supplier I.D. #	Supplier Name	# of Die Part Numbers & # of Dies Built			Average Delivery Time			Remarks
		MTD	YTD	History	MTD	YTD	History	

This report is automatically generated, although remarks can be added as desired.

Figure 9.5 Supplier Performance on Die Delivery Report.

to be automatically generated from data input into Figure 9.4, which would provide performance data for each of the company's die suppliers. This data could then be used to accurately evaluate and communicate with the various suppliers and to make better-informed sourcing decisions.

The Bottom Line

Again, the concepts of this system can be used universally when critical value streams (critical paths) can be identified. The system works because it is true that when you measure performance, performance improves. The system also provides the detailed information necessary to know where the biggest opportunity areas are and where increased attention should be focused. Of course, another big advantage of the system is that you have a permanent record of what was achieved on each program as pertaining to TPCT as well as other pertinent data that can be added to the formats.

Chapter 10

What Size Should the Cycle Quantity (Lot Size) Be?

Over the past several decades, there has been a gigantic shift in perspective, at least in many quarters, regarding the constitution of effective manufacturing processes. This has been largely brought about by the introduction of Lean Manufacturing and corresponding Lean principles. We have already reviewed several instances in which Packard, in an effort to become "leaner" through the introduction of Lean principles, actually did anything but accomplish that. As we discussed, based on my observations and suppositions, this occurred due to a lack of confidence on the part of key managers that they knew the right answers, an unwillingness on the part of many managers to challenge directives from top executives even when major disagreement existed, and a lack of understanding on the part of many executives of the consequences of making some of the bad decisions that were made (although the ramifications were generally very easy to predict).

My observations are that when the pendulum swings, it normally swings way too far in the opposite direction, until reason brings it back into equilibrium. A good example is how business abuses in the early part of the twentieth century brought about the unions. The pendulum has swung way past vertical over the past several decades to the point that unions now have way too much power and (along with weak and shortsighted management) have been a major factor in the demise of virtually every industry, business, or service which they touch. Just look at the steel industry, the auto industry, the airlines, public education, and government employees unions, to name a few. If the pendulum is not brought back to vertical soon in those industries (service, manufacturing, and government) where unionization is strong, performance will continue to decline and companies and entities will eventually dissolve or be forced to relocate (except, of course, for government entities, which will be a primary factor in the ultimate bankruptcy of this country if things don't change soon. According to CBO [Congressional Budget Office] numbers, federal employees with less than a bachelor's degree make substantially more than their private-sector peers on a per-hour basis [around 20%] and have benefit packages over 70% as generous. This wasn't the case even 10 years back when many individuals went to work for the government for job security, even though the pay and benefits were inferior to the private sector).

When I started with Packard in the early 1970s, batch processing was still a favored manufacturing process where it made sense, and no one was afraid to admit that batch processes were being utilized. This has changed over the past four decades at Packard. If you were courageous enough to say anything positive about batch processing within Packard since about the early 1990s, even though it was far-and-away the most effective manufacturing process for virtually all cutting and lead prep operations within the company (along with a lot of component assembly and final assembly), you would draw evil stares and nasty remarks. Some would tell anyone so bold to quit living in the past or quit being closed-minded and learn to accept and grow with the new realities. Or the statement I dislike the most: You have got to change paradigms.

In looking back, I think one of the biggest problems with the image of batch processing was the fact that it was *called* batch processing. If batch processing had been renamed Optimum Quantity Manufacturing (or OQM for short), or something similar, it probably would not have been so stigmatized and Packard might not have made a lot of the really bad decisions it made. (It's kind of like how we are being told by many in the media and the administration not to refer to acts of terror as "terrorist acts" anymore, in this politically correct world, but, we instead are supposed to refer to them as manmade disasters. They will always be terrorist acts in my world.)

In the early days, Packard knew and understood the value of batch processing, and, based on some evidence, actually tried to determine what the proper batch size should be in order to maximize the efficiency and cost effectiveness of a process. For example, for wire cutting, a column was included on the computerized cutting report called Planned Cut Quantity (PCQ). When the inventory of a lead (a cut and terminated wire) reached the reorder point, additional leads were pulled from the Cutting Department many Packard executives just never seemed to understand the fact that this was a pull system). The quantity that was cut by the Cutting Department was the PCQ, assuming that other urgent orders or shortages of materials did not interfere.

It is apparent that this PCQ was an attempt by someone, or some ones, at Packard to establish what I would later call the Optimum Cut Quantity. I was never able to discover who established this quantity or what rationale they used. In most cases, it appeared the PCQ was in the neighborhood of three weeks of inventory. I suspect this figure was reached because our customer authorized us to have three weeks of cut leads in the plant, so if any engineering change made a lead obsolete, GM would have financial responsibility for up to three weeks worth of inventory. The rest of the justification undoubtedly was due to the fact an average changeover time on a cutter was about 15 minutes, and Warren labor was very expensive—the fewer changeovers, the better (to reduce the cost of labor, tools, and equipment). Changeovers also generated scrap, and most quality problems were created due to an improper setup, so the thinking had to be: "Let's minimize setups."

Part of the reason that setups took so long (in addition to the fact that, to a large degree, it is the nature of the beast) was that there was no scientific analysis made to ensure that each lead was assigned to the optimum cutter and was grouped with other similar leads, which would be sequenced and cut in such a way that setup time would be minimized. In the early 1970s, computers and computer programs were not available that would have allowed this type of activity to be easily accomplished. My wonderment was that it was not even done several decades later, when computers could have been easily and effectively utilized, especially since we had implemented just such a system in South Korea.

In the mid-1990s, Packard reintroduced cutting and lead prep to the Mexican Operations after spending millions on remote cutting and lead prep. One of the Packard executives, who had a lot of manufacturing experience, was clever enough to realize that if he could come up with a cool sounding name with a pithy acronym, he could probably convince the top brass to adopt the "new" concept—which he did. The name of the new concept was the Strategic Manufacturing Initiative (or SMI for short), and a group of over 20 individuals was assigned to make the new initiative a reality in the Mexican Operations—for the second time.

The group did some good things. For example, they designed a new type of flow-through rack on which to hang and control cut leads. They also made the correct decision to load each lead on only one specific cutter and to try to group similar leads on each cutter to help minimize setup time (i.e., the more similar a lead was to the proceeding lead, the less complicated and time consuming the setup would be). Their failure was in not using the computer effectively to ensure 100% accuracy in loading each machine, and, even more importantly, there was a failure to use the computer to establish the perfect sequence in which to sequence and cut leads, which would help drive the average setup time to the absolute minimum.

These are the kinds of things that cry out for specialized computer programs, which will never be available commercially, but, fortunately, can be designed and developed relatively easily internally, and which really can drive improvement. However, the concept of using the computer to drive improvement seemed to be a foreign concept to most Packard executives. The SMI group did do a lot of work on Single Minute Exchange of Die (SMED) analysis and did make some improvements, which is a good thing. But a computerized cutter loading and sequencing system could have garnered significant additional savings with not that much effort or cost.

I had already designed a system to do these things when I was in South Korea. The system took me a couple of days to think through and design, and we were able to program it in a few weeks. Best of all, most of the data we needed already existed in the database. It functioned so beautifully and proved to be such an asset in helping us run an efficient cutting area that I couldn't understand why a similar program had not already been developed. I shared this system with several Packard executives who came to visit us in South Korea, and, although they were very impressed with the system and the efficiency of our cutting area, no efforts were made (as best I could tell) to determine if the system had value for the Warren plant (which it most definitely did) (Figures 10.5 and 10.6).

After I had returned to the States and SMI was initiated, I met with the new SMI manager briefly and reviewed with him our successful South Korean system. I also offered to help them implement something similar, if there was any interest on their part. It was pretty obvious that this new group wanted to put its own brand on what they developed, and there was no interest in borrowing a tool from someone else, regardless of how effective it was. I knew how to take a hint, so I just went about my business, still wondering how a company so big, and seemingly so successful, could consistently fail to take advantage of golden opportunities and make such bad decisions.

I kept track of what was transpiring within the SMI group and, as I observed what they were doing, an additional glaring failure became apparent. I assumed they would be coming up with some type of scientific means to answer the age old question of the proper cut quantities for the various leads on the various different types of cutters. In the old days, the answer was pretty much to cut whatever the customer would authorize, but this was largely based on opinions of what the right answer was as opposed to scientific analysis (although, they could be somewhat excused due to the state of computers at that time). In the 1990s, as SMI was being designed, most Packard managers took the default position that cut quantities should be as low as possible, maybe down to as little as one day's requirements or less (although some managers with actual cutting experience knew this was nuts). However, again, this position was strictly based on opinion and not on any scientific analysis, as Toyota said should be done when making any decision.

The bottom line is that no one really knew what the right answer was. All Packard had were a lot of opinions. But, I kept thinking that we did not have to guess at the answer, it should not be that difficult to come up with a computerized system to consider all of the various factors and make a determination based on real scientific analysis. I didn't develop my thoughts any further at that time; it wasn't my area of responsibility and they seemingly didn't want my help. And, besides, it was rather entertaining watching all of the various arguments shaping up, based on nothing but opinion and conjecture.

When I left the company in 2001, Packard still did not have any system in place to establish an optimum cut quantity (nor an optimum process quantity for anything else, for that matter) that would stand up under any kind of scientific scrutiny. Cut quantities varied from plant to plant, based on the experiences of the plant personnel and what their motivations were. One-day cutting cycles were tried in a plant or two, with the kind of results you might expect, so I think everyone had pretty much gotten off that bandwagon. However, the proper determination of the optimum cut quantity was still a mystery.

It shouldn't have been.

Portugal Will Show the Way

When I arrived in Portugal in January of 1984, their cutting system was not unlike the system that was in place in the Warren plant in the 1970s (which was not a bad thing). They did not have a computerized cutting system to load and sequence leads on the various cutters either, but they did seemingly do a pretty good job of manually loading cutters and grouping leads appropriately. They also had planned cut quantities in the neighborhood of two to three weeks.

The cutting area obviously ran fairly well because, in my responsibilities as engineering manager, I don't recall any significant problems in the wire cutting area. This was great because most of my time was occupied trying to get a new Methods Lab up and running and all of the engineers thoroughly trained.

Then, something happened one day that I knew could change all of that due to the tremendous havoc that would be created in the cutting area if it were implemented. A new general manager from Germany was assigned to Portugal, who clearly wanted to put his stamp on the Portuguese Operations and to make a name for himself. He had a lot of good ideas, but he also had some that were from left field. One of those was to arbitrarily reduce the cut quantities from two to three weeks down to one week or less. This was about the time that Lean Manufacturing was getting strong press, and he was going to be the first guy on the block to "get Lean" in the cutting department in the European Operations.

We had a couple of discussions in which I explained to him that I thought it would be a big mistake, to dramatically reduce our cutting cycle to one week, especially if we tried to convert the plant all at once. We would need several more cutters, more operators, and more maintenance people for those cutters; more floor space for those cutters that we did not have; and additional tools for those cutters. He countered that we would save a lot of inventory dollars and the lower cut quantities would somehow make us more efficient, although he had a real hard time explaining how that miracle would occur. It was obvious he was not going to be persuaded through normal opinionated discussions. He had his own opinions and they were as valid as mine—in his mind only, not in mine.

I knew I needed to try a different approach, one that used scientific reasoning and analysis with which he could find no fault. As I thought about it, it occurred to me that it would not be that difficult. When analyzing a cutting operation (or virtually any manufacturing process, for that manner), there are two "buckets" of cost factors that must be analyzed.

Cost Bucket One

The first "cost bucket" contains costs that decrease on a per unit basis (or on a yearly cost basis) as the process quantity (cycle quantity or cut quantity) is increased. In a Cutting Department, these would be costs associated with the number of cutting machines required; costs such as:

- Investment for cutters (opportunity cost, much like the cost of inventory)
- Investment for dies and other tools
- Depreciation
- Spare parts
- Maintenance
- Direct operators
- Indirect operators (quality operators, service operators, etc.)
- Supervision
- Floor space required for cutters
- Scrap generated at setup

To envision this concept, imagine a cut quantity of one lead. It would take roughly 10 minutes to complete a setup for this lead, then some scrap would be generated while qualifying the setup, then one second would be spent to cut one lead. The machine would then be stopped and set up for the next lead. This process would continue until all leads had been cycled. Then the process would repeat itself in order to supply all of the leads required for final assembly. In this scenario, the unit cost per lead (or maybe better understood, the yearly cutting cost for the department), for the above cost factors, would approach infinity. We would only have one second of productive labor for every 10 minutes of nonproductive labor. Even a child would understand that this would make no sense (so much for Toyota's vision of the ideal system of lot sizes of one, at least as understood by Packard).

If we increased the cut quantity to 100 pieces, the unit cost would come down dramatically, but it would still be very high. We would still only have 100 seconds of productive labor (less than 2 minutes) for every 10 or so minutes of nonproductive labor.

If we increased the cut quantity to 1000, we now have a little over 16 minutes of productive labor for every 10 or so minutes of nonproductive labor, which means that almost 38% of our labor is still nonproductive.

If we increase the cut quantity to 3000, the numbers are starting to look a little better, but we still have almost 17% of nonproductive time. If we increase the cut quantity to 6000, the nonproductive time drops to less than 10%.

If you charted a graph with unit cost (or yearly cutting cost) on the y-axis and cut quantity on the x-axis, the curve would start out approaching infinity with a cycle quantity of one; drop rapidly through a cycle quantity of a thousand or so (depending on the cost factors for the given operation, i.e., the cost of money, hourly labor cost, etc.); continue to drop, but not as rapidly through cycle quantities approaching 5000 or 6000; and then drop modestly until reaching cycle quantities of 15,000 or so, at which time the curve almost starts to flat line as larger cut quantities would add little additional productive time on a percentage basis.

Cost Bucket Two

The second "cost bucket" contains costs that will increase on a per unit basis (or a yearly cost basis) as the cut quantity increases. These costs are associated with inventory; costs such as:

- Inventory cost (actually the carrying costs of inventory and not the total value of the inventory)
- Inventory management costs
- Floor space devoted to inventory
- Racks and totes required to house inventory
- Potential excess and obsolete material

At very low cut quantities, the unit cost (or yearly Cutting Department cost) for these cost factors is extremely low. At a cut quantity of one piece, the costs would be virtually zero for inventory, floor space required, and potential excess and obsolete, although the storage and handling of the materials would be a bit of a challenge. As cycle quantities increase along the x-axis, the costs for the above cost factors also would increase; however, the costs would increase slowly (on much more of a linear basis, as opposed to the exponential decrease for the costs in Cost Bucket 1 as cycle quantities were increased).

Now, how do we use these two curves depicting the actual costs that are contained in the two cost buckets for the Cutting Department? It is actually quite simple. For each of the cut quantity volumes shown on the x-axis, all we have to do is to add the corresponding unit costs (or departmental yearly cutting costs) from the two curves (one representing bucket one costs and the other representing bucket two costs). The resulting total cost curve will track the cost bucket one curve, while staying above it, with increased separation with increasing cycle quantities, until the two curves intersect (i.e., the Cost Bucket 1 and Cost Bucket 2 curves intersect). The combined curve will then start to track the Cost Bucket 2 curve, with decreasing separation as the cycle quantities increase.

The optimum cut quantity will be approximately the point at which curve one and curve two intersect. In some cases, the yearly cutting cost increase to move somewhat left on the x-axis is not gigantic, because the total cost curve may start to flatten out before reaching the intersection point. If so, a decision can be made to reduce the cut quantity to something less than the optimum without too much of a penalty. However, by using this kind of scientific analysis, it will be known what the cost penalty will be (Figure 10.1).

In today's world of technology, it is relatively easy to do a good job of making this type of analysis on an Excel® spreadsheet. Unfortunately, Excel wasn't released for the PC until 1987, so I decided to do a quick and dirty analysis that would provide a ballpark answer, since this concept was on the front burner and I needed to come up with something fast to kill our new general manager's one-week cutting cycle dream before it was too late and we suffered the negative consequences. In cost bucket one, I considered the investment for cutting machines, the additional floor space that would be needed for cutting machines, and the additional direct labor that would be needed to run the machines. In cost bucket two, I considered the cost of inventory and the floor space requirements that would be needed to support this inventory. I figured the other costs, which were not considered in the calculations, would wash each other out. At least, I figured it would give us a quick answer that was "good enough for government work," and certainly better than the opinions that were being freely tossed around.

Based on my analysis, the optimum cut quantity, which provided the lowest yearly Cutting Department cost, was between two and three weeks based on our plant volumes. This was not a case of knowing the answer and working backward. These calculations were based on the fact that Portuguese labor rates were low and interest rates were high; both of these factors favor a smaller cut quantity. I plugged in the factors for Germany, where labor costs were higher and interest rates lower, and the optimum cut quantity was in excess of three weeks, based on their average volumes.

Some People Don't Want to Be Confused with Facts

Once I had this analysis developed, I went in to see the general manager and told him the good news. I don't remember him taking the news all that well. His comments were along the lines of: "I don't care what the numbers

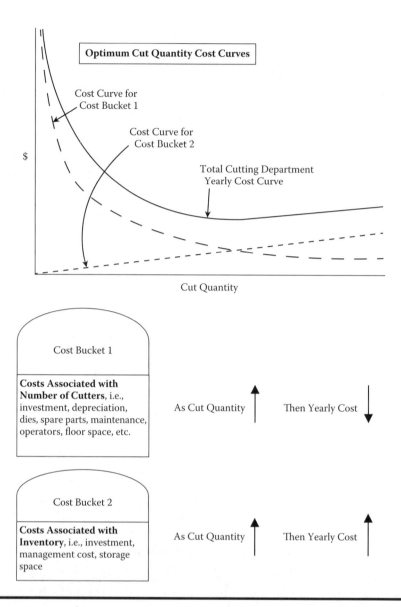

Figure 10.1 Cost Bucket 1, Cost Bucket 2, and Total Cost Curves for a typical cutting department.

say, I believe my idea is a good one and something we need to do, so we are going to do it." It was clear to me that he had read one too many books on the new concepts of Lean sweeping the business world, and there was no way I was going to dissuade him.

As fortune would have it, there was a Packard Europe conference scheduled a couple of weeks after I had this sit-down with the general manager and he had been asked to give an update on the Portuguese

Operations, especially regarding any new concepts that had been implemented or that were being considered. He made it clear to me that he intended to present his concept of reducing cutting cycle quantities (by 50 to 70%) in this meeting, so that he could get the European leadership on board (and, I'm sure, to guarantee their approval of the new funding when he came asking for it). I reviewed the numbers with him again and told him I thought it would be a mistake to present this concept in the meeting. There was just no way we could justify his project. We would need to submit a request for several additional cutters (each costing almost $100,000), and there was just no way to justify them. But, he made it clear he was going to proceed.

The trip to Dublin, Ireland, proved to be a fantastic experience. I got to see the raw beauty of Ireland and the quaintness of Irish towns, a chance to play the world renowned Portmarnock golf course, and the opportunity to have my faith in at least some Packard executives restored.

When it was time for my boss to present, I held my breath. Was he still going to present his new concept and have his head handed to him, or, worse yet, receive a standing ovation and be told to proceed at full speed? The answer turned out to be the former, fortunately, although I felt a little sorry for my boss.

He presented his concept and extolled the virtues of it, most especially the significant reduction in inventory. He presented no facts or figures, just the notion that reducing inventory would more than offset any other contradicting factors. When he had completed his presentation, the engineering director for European Operations remarked (and I paraphrase), "I see that we would be reducing inventory, and that is a good thing, but wouldn't we need to buy a lot of additional cutting machines and hire a lot of additional operators to run them?"

My boss was like a deer caught in the headlights. He surely should have expected such a question, but apparently he hadn't. I guess he had sat in a lot of the same types of meetings I had over the years, in which someone presented something that defied logic and common sense, but who was not contradicted or challenged due to some sort of unwritten business etiquette. This was a different type of leader—one that I truly admired. My boss stammered around for a minute or two, but obviously he did not have a good answer to the very fair and pertinent question that had been posed. In the end, the engineering director said (again paraphrasing), "It is an interesting concept, but we need to do a lot of analysis before we decide that this is something that is in the best interests of our company."

I spent another year in Portugal before being asked to take the Joint Representative Director assignment in South Korea for the new Packard joint venture (Shin Sung Packard). Nothing more was mentioned about reducing cut quantities during that year. The concept had died in a hotel conference room in Dublin, Ireland—thank goodness.

However, since the Methods Lab was up and running and all of the engineers had been thoroughly trained, I had a little more time available to look into other things. Because so many of Packard's investment dollars were tied up in cutting machines, I decided to take a more intense look into our cutting practices and systems. One critical factor that prohibited us from justifying a reduction in our cutting cycles was the relatively lengthy, average changeover time we were experiencing, probably somewhere in the 12- to 15-minute range, depending on how well the cutters had been loaded. A lot of changeovers meant a lot of nonproductive setup time.

A lot of time had been spent by Packard, in all of its operations, to try to make setups faster, and there wasn't much more that could be done with current technology. However, there was one glaring omission, something that had not been done that could save a lot of time in making setups, if it were implemented. That glaring omission was in not using the computer to optimally load and sequence leads on each of the cutters.

A setup could be comprised of one or more of the following: a change of the cut length, a change of the wire type, a change in strip length on one or both ends, a change of the terminal application die on one or both ends, and a change of terminal on one or both ends; and, in cases where seals were applied to one or both ends, or in cases where doubling applications would be made on the cutters, additional setup steps would be required. Each of these setup steps required a minimum amount of time to accomplish and these times would vary based on cutter type, tool type, and material type.

I reasoned that if leads could be loaded and sequenced perfectly on each cutter by the computer, a tremendous amount of setup time could be eliminated, which would thereby change the economics such that we could justify reducing the cut quantities to some degree.

Free Rein in South Korea

I did a lot of thinking about how to accomplish this perfect loading and sequencing while still in Portugal. The general manager thought it would be a great idea to optimally load and sequence each cutting machine, because he believed it would help him achieve his desire to reduce cutting cycles. He suggested we could determine the perfect sequence by determining the precise times to accomplish each of the potential steps in a setup, pick a lead, and then determine which lead, of all of the other leads in the plant, would require the least amount of time to set up as compared to the first lead, and then determine which of all other leads in the plant would require the least amount of time to set up as compared to the second lead, and so forth. I knew enough about programming to realize that this would be a programming nightmare and would take forever to run if it ever were successively programmed. However, after thinking about it for a while, I realized there was a fairly simple way to accomplish the task, which would be relatively easy to program and would not require a lot of time to run once the programming was complete.

At about the time that I had figured out how to design a system to optimally load and sequence leads on each cutter, I was asked to take the South Korean assignment. I explained to the Portuguese general manager how the proposed system would work, but after I left and there was no one there to spearhead the project, I think it died a rather quick death. However, I knew this system I designed would be one of the first I would implement in South Korea. I was confident it could significantly improve our cutting productivity versus the normal way cutting departments were organized and run. And I knew I would have the computer flexibility and authority to make it a reality.

In a few days I designed the program complete with all input screens and reports, and our MIS Tech had it programmed in a few weeks. This program worked exceptionally well in South Korea and enabled us to run much higher levels of productivity in the cutting area than were being run in other areas of Packard. It is a prime example of how the computer can be used to drive improvement; in this case, improvement not only in productivity, but also quality, as the setups were greatly minimized and simplified (Figures 10.6 and 10.7).

This is the type of program that can give a company a competitive advantage; the type of computer program that is not available on anyone's shelf. It is a program specific to a wiring harness cutting department and designed by a knowledgeable user in that department. There is just no way a computer programmer could be expected to understand the cutting process well enough to design such a system (along with the complementary lead prep system that I also developed).

We got this program up and running in a relatively short period of time and with minimal expense. It certainly paid for itself a thousands of times over. For companies big enough to justify an MIS Department, it makes all the sense in the world to design and implement specialized programs, such as this, with internal resources. It will save time and money (including ongoing maintenance costs) as well as protect technology. Smaller companies, or those without MIS capabilities, may be able to contract services for less than "an arm and a leg," assuming a desired system is well defined and detailed (to include input screens and reports). However, my experience is that it is best to do it internally, if possible.

Cutting machines are not the only pieces of equipment on which proper sequencing is an important factor. Certainly, every reasonable step should be taken to increase uptime on a piece of equipment (especially a high investment piece of equipment) by having all materials onsite when the changeover is to be made and by analyzing each step in the setup to determine opportunities for improvement. However, one of the easiest ways to drive improvement is to sequence work such that setup time is minimized.

Determining Optimum Cut Quantity (or Optimum Process Quantity)

After I left Packard, I consulted with another big automotive component supplier for a period of several years. One of their businesses was automotive wiring; a business they were trying to grow. One of their key operations was located in Cebu, Philippines and they were having some difficulty meeting their customer requirements due to some ongoing problems with productivity and quality. I was asked to spend three months helping the Philippine Operations get back on the right track, especially in regards to Industrial Engineering (IE) controls. The stateside managing director had some prior experience with Packard, and he knew that the IE systems that were in place in the 1970s at Packard could be a big help to their Philippine Operations (little did he know what a small portion was still left in the "new" Packard).

I spent a couple of months putting several IE systems and controls in place, including a Methods Lab, so that they could start preplanning their new products (and redo the manufacturing systems for their existing products, as time permitted), and thereby avoid many of the problems they had been having. But, after spending two months in their various plants, it was obvious that they had other major problems, especially in their wire cutting departments. Problems in the cutting areas were creating as many, if not more, of the issues with productivity, delivery, and quality as poor engineering.

After I had trained all of the engineering supervision and the Methods Lab was up and running, I decided to take a much closer look at what was going on in the cutting area. Based on my prior experiences, I thought there were a few simple concepts and/or tools that I could help them put into place that might make a big difference in their performance.

On closer review, there were four major problems that needed to be corrected:

1. Leads were not assigned to specific cutters, but were cut on whichever cutter was available when a lead needed to be cut.
2. From no. 1, it was obvious that no effort was made to sequence leads in such a way as to minimize setup time.
3. Visual control was virtually impossible with the type of cut lead storage system that was being used.
4. And, maybe most importantly, they were trying to cut less than a one-week cycle quantity (sometimes as little as a few hundred leads before making a changeover) due to a company directive.

In the years since, I had tried to convince the Portuguese general manager of his error (through the use of manual calculations) in making a decision to arbitrarily reduce the cut quantity to one week or less, computer technology had made great advances and the Excel spreadsheet became a reality. Based on my recent Packard experience of having one of my engineers use Excel to automate various forms used in the preplanning process, I knew that this tool could be used to make a scientific analysis regarding the optimum cut quantity (or the optimum process quantity), which was much more thorough and accurate than what could be done manually.

I got some help from one of the Philippine engineers, who had experience with Excel, and I designed and she programmed an Excel spreadsheet in a matter of a few hours. The system was designed using the concepts I explained earlier, but because Excel was available to make all of the calculations, I was able to consider all of the costs in both Cost Bucket 1 (costs related to the number of cutting machines needed; costs that would decrease as cut quantity increased) and Cost Bucket 2 (costs associated with the inventory of cut leads; costs that would increase as cut quantity increased) (Figures 10.2, 10.3, and 10.4).

Once the system was designed and programmed, I reviewed it with the engineering manager and his supervisors responsible for Industrial

Optimum Cut Quantity (OCQ) System Formats

| Location | Mexico | Cutting Plant | Alpha | Cutting Dept. | 101 | FA Dept. | All | Cutter Type | All | Customer(s) | Gm, DCX |

Package(s)

Assumptions

Work days per year	240	days	Terminal die cost	$2,500	per die
# of circuit codes	3,396	different codes	Terminal die depreciation rate	33.3%	per year
Total circuits to be cut per day	1,200,000	circuits per day	Terminal die maintenance	$120	per cie per year
Avg. daily volume per harness family	353.4	Avg. harnesses per day	Perishable tool cost for a terminal die	$437	per cie per year
Planned cutter capacity	110%		Seal applicator cost	$4,000	per seal applicator
Cost of a cutter	$110,000	per cutter	Seal applicator depreciation rate	33.3%	per year
Cost of investment money	5.0%	per year	Seal applicator maintenance	$150	per seal applicator per year
Cutter depreciation rate	14.3%	per year	Perishable tool cost for a seal applicator	$200	per seal applicator per year
Maintenance cost per cutter	$13,500				
Avg. circuits cut per minute of run time	50.0	circuits per minute	Floor space requirement per cutter	35.0	sq. meters per cutter
Average setup time	9.0	minutes	Yearly cost per sq. meter	$400	per sq. meter
Average scrap rate per set-up	10.0	meters per set-up	Safety Stock in days	1.0	days
Seal application: Yes or No	NO		Average individual circuit cost	$0.30	per circuit
			Average inventory cost for circuits	$360,000	for 1 day of inventory
Cutter scrap cost	$0.15	per meter	Inventory carrying cost	5.5%	per year
			Floor space required for wire storage rack section	1.0	sq. meters per rack section
Avg. work minutes per indirect employee	529	work minutes per indirect employee	Cost of a rack section	$700	per section including hooks
# of direct operators per cutter	1.17	operators per cutter / shift	Cost for a cutter operator per year	$15,000	per operator per year
Minutes/setup for Quality Indirect	10.00	minutes / setup	Cost for a quality indirect/year	$10,000	
Minutes/setup for Materials Indirect	7.00	" / setup	Cost for a materials indirect/year	$10,000	
Minutes / setup for Crib Indirect	12.00	" / setup	Cost for a crib indirect / year	$10,000	
Minutes / setup for Warehouse Indirect	5.00	" / setup	Cost for a warehouse indirect / year	$10,000	
# of cutters per Team Leader	20	cutters per person / shift	Cost for a team leader / year	$20,000	
# of " per salary personnel	20	" per person / shift	Cost for a salary person / year	$40,000	
Cutting shifts available	2.00	shifts per day			
Paid minutes on AM shift	565	minutes	3rd Shift Burden Add	$150,000	$ per year over and above
Paid minutes on PM shift	525	"			
Paid minutes on 3rd shift	325	"	Allowable fixed delay percentage (excludes variable delays and set-up)	3.0%	Fixed Delays
3rd Shift loading factor	0%		Paid minutes per day	1,090	minutes
Effective paid minutes on 3rd shift	0	minutes	Allowable fixed delay time excluding set-up (*minutes*)	33	minutes

Figure 10.2 Cutting Department data for the OCQ System—the bold numbers are automatically calculated from the input data.

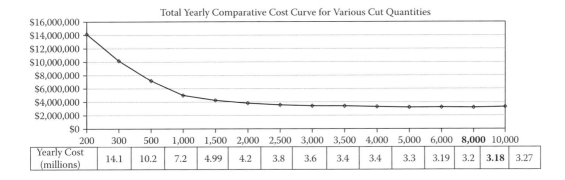

Figure 10.3   Yearly Cutting Department Cost Curve for various Cut Quantities—the bold numbers correspond to the Optimum Cut Quantity and the Total Departmental yearly cost that this OCQ would generate.

Engineering, Process Engineering, and Plant Engineering. They would need to understand the system and buy into it, and I also needed their help in obtaining the data that would be used to make the calculations; information such as the cost of a cutting machine, costs for dies and perishable tools, average cycle time to cut a circuit by cutter type, average setup time, average scrap rate per setup, cost of scrap, paid minutes per shift, cost per hour for various indirect operators, cost per hour for direct operators, average individual circuit cost, cost of a storage rack, cost of floor space, cost of money, depreciation rate, number of different lead codes, and number of leads required to be cut per day.

It took a couple of days to gather all of this data and, when it was all available, it was input into the Excel spreadsheet. The answer was very much as I expected. The Optimum Cut Quantity (OCQ) for each of the three plants (for their respective datum) was somewhere in the 3000- to 6000-piece range. However, the cost curve was relatively flat in this area, so some flexibility existed to establish cut quantities that were convenient for each plant and that did not create a risk of excess and obsolete material. For the higher volume packages, this number of pieces generally corresponded to two or three weeks of material.

The operations people were quite happy to have an objective tool to establish the optimum process quantities, quantities based on scientific analysis and not someone's opinion. They believed the tool could be used to convince their superiors that the minimum cut quantities, which had been dictated, were not justifiable. I recommended that they make a determination of how many days of inventory would be cut on each setup for the higher volume packages, and then cut the same number of days of inventory for each lead, even for low-volume packages. This would allow them to

Engineer J.B. Tryin

OPTIMUM CUT QUANTITY Report

Page 2 of 2
Date 4/23/04

Area of Consideration w/Cost	200	300	500	1,000	1,500	2,000	2,500	3,000	3,500	4,000	5,000	6,000	8,000	10,000
Planned cut quantity (circuits / set-up)	0.6	0.8	1.4	2.8	4.2	5.7	7.1	8.5	9.9	11.3	14.2	17.0	22.6	28.3
Planned cut quantity in days	20.0%	17.0%	14.0%	12.0%	10.0%	8.0%	7.0%	6.5%	6.0%	5.5%	5.0%	4.5%	4.0%	3.5%
Variable Delay Allowance	839	872	905	927	948	970	981	986	992	997	1,003	1,008	1,014	1,019
Average cutting & set-up time available (minutes)	$0	$0	$0	$0	$0	$0	$0	$0	$0	$0	$0	$0	$0	$0
3rd Shift Burden Add	4	6	10	20	30	40	50	50	70	80	100	120	160	200
Average raw cutting time / cycle (minutes)	13	15	19	29	39	49	59	69	79	89	109	129	169	209
Average cycle time including set-up (minutes)	64.6	58.1	47.6	31.9	24.3	19.8	16.6	14.1	12.6	11.3	9.2	7.8	6.0	4.9
# of set-ups per day per cutter	12,912	17,440	23,808	31,948	36,473	39,596	41,568	42,889	43,945	44,825	46,000	46,895	47,986	48,763
Daily capacity per cutter	92.9	68.8	50.4	37.6	32.9	30.3	28.9	28.0	27.3	26.8	26.1	25.6	25.0	24.6
# of cutters required @ 100% capacity	102.2	75.7	55.4	41.3	36.2	33.3	31.8	30.8	30.0	29.4	28.7	28.1	27.5	27.1
# of cutters required @ planned capacity	103	76	56	42	37	34	33	31	31	30	29	29	28	28
Actual cutters installed	110.8%	110.5%	111.1%	111.8%	112.5%	110.8%	110.8%	110.8%	113.5%	112.1%	111.2%	113.3%	112.0%	113.8%
Actual cutting capacity	32.0	43.7	60.6	80.9	91.8	99.9	106.1	109.5	109.5	111.2	117.3	117.1	121.3	121.3
# of circuits assigned per cutter	646	581	476	319	243	198	166	143	126	112	92	78	60	49
Meters of scrap wire per day	228	309	168	126	111	102	96	93	90	87	84	84	84	84
Scrap cost per day	$23,242	$20,928	$17,142	$11,501	$8,754	$7,127	$5,986	$5,167	$4,520	$4,034	$3,312	$2,814	$2,159	$1,755
Cutter investment	$11,330,000	$8,360,000	$6,160,000	$4,620,000	$4,070,000	$3,520,000	$3,410,000	$3,410,000	$3,410,000	$3,300,000	$3,190,000	$3,190,000	$3,080,000	$3,080,000
Cost of cutter investment money per year	$566,500	$418,000	$308,000	$231,000	$203,500	$187,000	$176,000	$170,500	$170,500	$165,000	$159,500	$159,500	$154,000	$154,000
Cutter depreciation per year	$1,613,604	$1,194,310	$880,018	$660,013	$581,440	$581,440	$502,867	$487,153	$487,153	$471,438	$455,723	$455,723	$440,009	$440,009
Cutter maintenance cost per year	$1,399,500	$1,026,000	$756,000	$567,000	$499,500	$459,000	$432,000	$418,500	$418,500	$405,000	$391,500	$391,500	$378,000	$378,000
Perishable tool cost per year for a terminal die	$135,033	$99,636	$73,416	$55,062	$48,507	$44,574	$41,952	$40,641	$40,641	$39,330	$38,019	$38,019	$36,708	$36,708
Estimated die required	309	228	168	126	111	102	96	93	90	87	84	84	84	84
Die investment	$772,500	$570,000	$420,000	$315,000	$277,500	$255,000	$240,000	$232,500	$232,500	$225,000	$217,500	$217,500	$210,000	$210,000
Cost of die investment money per year	$38,625	$28,500	$21,000	$15,750	$13,875	$12,750	$12,000	$11,625	$11,625	$11,250	$10,875	$10,875	$10,500	$10,500
Die depreciation per year	$257,497	$189,998	$139,999	$105,499	$92,499	$85,499	$79,999	$77,499	$77,499	$74,999	$72,499	$72,499	$69,999	$69,999
Seal applicator maintenance cost per year	$37,080	$27,360	$20,160	$15,120	$13,320	$12,240	$11,520	$11,160	$11,160	$10,800	$10,440	$10,440	$10,080	$10,080
Estimated seal applicators required	155	114	84	63	56	51	48	47	47	45	44	44	42	42
Seal applicator investment	$	$	$	$	$	$	$	$	$	$	$	$	$	$
Cost of seal applicator investment money per year	$	$	$	$	$	$	$	$	$	$	$	$	$	$
Seal applicator depreciation per year	$	$	$	$	$	$	$	$	$	$	$	$	$	$
Seal applicator maintenance cost per year	$	$	$	$	$	$	$	$	$	$	$	$	$	$
Perishable tool cost per year for a seal applicator	3,605	2,660	1,960	1,470	1,295	1,190	1,120	1,085	1,085	1,050	1,015	1,015	980	980
Floor space required for cutters (m²)	$1,442,000	$1,064,000	$784,000	$588,000	$518,000	$476,000	$448,000	$434,000	$434,000	$420,000	$406,000	$406,000	$392,000	$392,000
Cutter floor space cost per year	241	178	131	98	87	80	75	73	73	70	68	68	66	66
Direct heads required for cutting	$3,615,300	$2,667,600	$1,965,600	$1,674,200	$1,298,700	$1,193,400	$1,122,300	$1,088,100	$1,088,100	$1,053,000	$1,017,900	$1,017,900	$982,800	$982,800
Total cutting labor cost per year	125.8	83.6	50.4	25.4	17.0	12.7	10.1	8.4	7.4	6.4	5.0	4.3	3.2	2.6
Quality Indirect Labor Cost	$1,257,891	$835,759	$504,395	$253,822	$170,182	$127,380	$100,647	$83,834	$73,627	$63,593	$50,468	$42,875	$31,770	$25,827
Indirect Quality heads required	88.1	58.5	35.3	17.8	11.9	8.9	7.0	5.9	5.2	4.5	3.5	3.0	2.2	1.8
Indirect Materials heads required	$880,523	$585,017	$353,077	$177,675	$119,128	$89,131	$70,453	$58,684	$51,539	$44,515	$35,328	$30,013	$22,239	$18,079
Materials Indirect Labor Cost	150.9	100.3	60.5	30.5	20.4	15.3	12.1	10.1	8.8	7.6	6.1	5.1	3.8	3.1
Indirect Crib heads required	$1,509,469	$1,002,887	$605,274	$304,586	$204,219	$152,796	$120,776	$100,601	$88,352	$76,312	$60,562	$51,450	$38,124	$30,993
Crib Indirect Labor Cost	62.9	41.8	25.2	12.7	8.5	6.4	5.0	4.2	3.7	3.2	2.5	2.1	1.6	1.3
Indirect Warehouse heads required	$628,945	$417,869	$252,198	$126,911	$85,091	$63,665	$50,323	$41,917	$36,813	$31,797	$25,234	$21,438	$15,885	$12,914
Warehouse Indirect Labor Cost	10.3	7.6	5.6	4.2	3.7	3.4	3.2	3.1	3.1	3.0	2.9	2.9	2.8	2.8
Team Leaders required	$206,000	$152,000	$112,000	$84,000	$74,000	$68,000	$64,000	$62,000	$62,000	$60,000	$58,000	$58,000	$56,000	$56,000
Team Leader Labor Cost	10.3	7.6	5.6	4.2	3.7	3.4	3.2	3.1	3.1	3.0	2.9	2.9	2.8	2.8
Salary personnel required	$412,000	$304,000	$224,000	$168,000	$148,000	$136,000	$128,000	$124,000	$124,000	$120,000	$116,000	$116,000	$112,000	$112,000

Figure 10.4 Yearly Cutting Department Cost Data for the OCQ System. All of these numbers are automatically calculated from the data (assumptions) from Figure 10.2. Light gray numbers are included in the total yearly cost calculation. The dark gray numbers are of particular importance to the Cutting Department team to aid in the proper organization and set-up of the department.

Engineer L.B. Tryin

OPTIMUM CUT QUANTITY Report

Page 2 of 2
Date 4/23/04

	1.3	1.4	1.7	2.5	3.1	3.8	4.5	5.2	6.0	6.7	8.1	9.5	12.3	15.2
Average days of cut wire inventory														
Total circuit inventory value	$461,880	$512,820	$614,700	$869,400	$1,124,100	$1,378,800	$1,633,500	$1,888,200	$2,142,900	$2,397,600	$2,907,000	$3,416,200	$4,435,200	$5,454,000
Inventory cost per year	$25,403	$28,205	$33,809	$47,817	$61,826	$75,834	$89,843	$103,851	$117,860	$131,868	$159,885	$187,902	$243,936	$299,970
Circuit codes per storage rack section	14	14	14	14	14	14	14	14	14	14	14	12	8	6
# of storage racks required	243	243	243	243	243	243	243	243	243	243	283	340	425	566
Cost of storage racks	$169,800	$169,800	$169,800	$169,800	$169,800	$169,800	$169,800	$169,800	$169,800	$169,800	$198,100	$237,720	$297,150	$396,200
Cost of rack investment money per year	$8,490	$8,490	$8,490	$8,490	$8,490	$8,490	$8,490	$8,490	$8,490	$8,490	$9,905	$11,886	$14,858	$19,810
Storage rack floor space required (m²)	243	243	243	243	243	243	243	243	243	243	283	340	425	566
Yearly wire storage floor space cost per year	$97,029	$97,029	$97,029	$97,029	$97,029	$97,029	$97,029	$97,029	$97,029	$97,029	$113,200	$135,840	$169,800	$226,400
TOTAL YEARLY COMPARATIVE COST	$14,150,131	$10,167,567	$7,155,604	$4,990,974	$4,246,059	$3,829,662	$3,563,083	$3,424,729	$3,403,406	$3,288,455	$3,194,350	$3,220,675	$3,180,866	$3,277,845
Planned cut quantity	200	300	500	1,000	1,500	2,000	2,500	3,000	3,500	4,000	5,000	6,000	8,000	10,000
Planned cut quantity in days	0.57	0.85	1.42	2.83	4.25	5.66	7.08	8.49	9.91	11.32	14.15	16.98	22.64	28.30
Total set-up time per day (minutes)	581	523	429	288	219	178	150	129	113	101	83	70	54	44
Total delays excluding set-ups per day (minutes)	251	218	185	164	142	120	109	104	98	93	87	82	76	71
Raw cutting time per day (minutes)	258	349	476	639	729	792	831	858	879	896	920	938	960	975
Raw cutting time as a % of paid time	23.7%	32.0%	43.7%	58.6%	66.9%	72.7%	76.3%	78.7%	80.6%	82.2%	84.4%	86.0%	88.0%	89.5%
Delay Factor (%)	29.9%	25.0%	20.5%	17.6%	14.9%	12.4%	11.1%	10.5%	9.9%	9.3%	8.7%	8.1%	7.5%	7.0%

Figure 10.4 (Continued)

Note: If desired, the system has a programming option that will prevent a high volume die from being split between two cutters, assuming the capacity exists on one cutter to manage the entire volume. This can also help spread out the lower volume applications among the cutters.

Optimum Cutter Loading & Sequencing

Cutter Loading

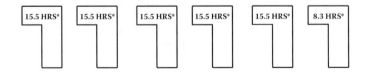

Hours/Cutter = Σ Volume/day × [Base Standard × (1+ Delay Factor % less Setup %) + Actual Setup (*in min/pc*)] ÷ 60 minutes/hour

* **Based on specific setup times and other standard delays considering the OCQ.** (*This example is based on a two shift cutting operation with 17 paid hours per day and with a capacity planning of 110% per machine.*)

Cutter Sequencing

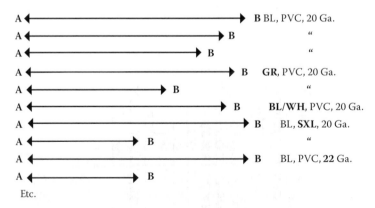

A = Strip **A**, Die **A**, Terminal **A**
B = Strip **B**, Die **B**, Terminal **B**

Lower to higher sequencing priorities: length, color, type, gauge, right end terminal, left end terminal, right end die, right end strip, left end die, left end strip, double terminal, right end seal, left end seal, right end seal applicator, left end seal applicator,

Figure 10.5 Simple Example of Cutter Loading and Cutter Sequencing.

cycle through each of the leads at approximately the same rate on each cutter, so that once leads were loaded onto specific cutters in the optimum sequence in order to minimize setup time, they could cut leads in that sequence with little risk of interruption. I also intended to help them implement the optimum cutting sequence system.

In addition to giving them a tool to establish the OCQ, I designed some special flow-through racks that could be used for visual control. I explained how leads should be loaded onto specific cutters, and how eventually each lead on each cutter should be sequenced in order to minimize setup time. I laid out the sequencing system for them, but I knew they would not have it programmed before I left.

Over the next three weeks, before I left the Philippines, the plants had implemented the new OCQs, installed some of the new racks, and manually loaded some of the cutters based on the criteria I had given them. During this short period of time, they went from a situation where they were working excessive overtime and still not meeting the requirements to a situation where they were cutting back on overtime and building up inventories.

I received a letter from the engineering manager about six weeks later. In part, the letter said:

> I wanted to give you an update on the process you initiated here in the Philippines. To date we have freed up 22 cutters; 8 we will retain for future business and 14 can be sold—a very significant cost savings. We have the system kicking out the sequencing data; this system needs ongoing maintenance (which is true of all computer systems; another great reason to program in-house if possible), but it is a great system. ... We are approaching 80% machine utilization, up from 40% when we started, and I anticipate further open machines. We have slashed appropriations on upcoming programs by well over a million dollars that was originally in them for new cutting machines.

Summary

These savings were generated through the use of computer programs that were designed to drive improvement (utilizing scientific analysis, which Toyota says should always be used when making decisions), coupled with a belief in the programs and a desire to fix some serious problems on the part of some dedicated employees. While the Optimum Cut Quantity program and the Optimum Sequencing program were designed specifically to optimize wire cutting for automotive wiring harnesses, the exact same concepts can be used in virtually any other process in which determining the optimum cycle quantity and/or determining the optimum sequencing can help drive down costs. At the same time, these tools (especially the OCQ program) can help eliminate the tendency of executives to use opinions to dictate policy, rather than let the facts speak for themselves.

Later in this chapter you will find sections from the IE Manual that I wrote concerning the Optimum Cut Quantity Subsystem and the Optimum Cutter Loading and Sequencing Subsystem. The OCQ data are basically complete as shown and programmed on Excel. It allows the user to change the data, based on changing conditions or when making evaluations for different plants or for various proposals, e.g., you can easily evaluate a three-shift versus a two-shift operation (see sidebar).

Surprising to many people in Mexico, a two shift cutting operation is cheaper and more flexible than a three shift operation when based on Mexican economics, although the results could be different for other locations, although probably not.

This system could obviously be put on a mainframe computer, but Excel provides an accurate answer at a very low cost and with a lot of flexibility and simplicity of use.

An explanation of the Optimum Cutter Loading and Sequencing Subsystem (OCLASS) also is attached, along with a sample of the Sequencing Report (Figure 10.6) and the Inspection Report (Figure 10.7), which are automatically generated for each cutting machine. There are several other reports I could have included, which provide performance information that is very helpful in running an effective cutting department, but these reports (along with the Optimum Cut Quantity report) are the primary computer tools that can be used to dramatically improve performance in a cutting area, and do so in a scientific manner.

Optimum Cut Quantity Subsystem

Of all factors that determine the total cutting cost within an operation, one of the most important is the established cut quantity. Unbelievably, most wiring harness companies have no scientific basis whatsoever to determine

What Size Should the Cycle Quantity (Lot Size) Be?

Cutting Inspection Report

Foreman _____ Plant _____ Cutter type _____
Location _____ Cutter # _____
Cutting Dept. _____
Customer(s) _____
Program(s) _____

#	Circuit #	Gauge	Insulation Type	Color	Circuit Diameter	# of Strands	Diameter of Strands	Seal Sensor L	Seal Sensor R	Visual L	Visual R	CFM L	CFM R	Visual L	Visual R	Strip Sensor L	Strip Sensor R	Left Strip	Right Strip	Center Strip Loc	Center Strip Dim	Circuit Length	Comments	Core CH Min.	Core CH Max.	FP#1	FP#2	#3	LP	Core CW Min.	Core CW Max.	FP	LP
1	PM1	0.75	SXL	BR	X	7	Z											6	4			1070		1.88	1.90					2.83	2.87		
2	PU1			WH/BR																			1230											
3	MF1			WH/GR																			680											
4	28		PVC	BK	XX																	610											
5	29X			GN															5															
6	P1		SXL	GR/WH	X																	1140											
7	75			BL																6			670											
8	25			BR																			970											
9	XM1		PVC	GN	XX																	610											
10	135		SXL	GR/WH	X																	1040											
11	24			WH/GN																			670											
12	F18		PVC		XX																	610											
13	152B	0.5		BK	YY		ZZ															690		1.60	1.62								
14	8E			GR															12				1030											
15	RU16																						1000											
16	FB4A	0.75	SXL	RD/WH	X		Z															920		1.88	1.90								
17	33A		PVC	WH/BR	XX																	810											
18	237			YL															5				1030											
19	PM4			BR																			810											
20	72A			WH/BR																			2300											
21	P4																																	
22	X70A	0.5		BK																6			680		1.60	1.62								
23	153BY				YY		ZZ															620											
24	153B																						520											
25	X70A																																	
26																																		
27																																		
28																																		
29																																		
30																																		
31																																		
32																																		
33																																		
34																																		
35																																		

Note 1: For doubles, use only one number in the first column, but use two rows for all of the remaining data, i.e., one row for each of the leads in the double.

Note 2: Everything in yellow is automatically populated from the computerized cutting system, which provides all of the specifications for each of the circuits to be cut on each specific cutting machine. This sheet is to be used to do complete first piece and last piece sample in inspections for all circuits assigned to each cutter. The operator is to fill in each block.

> If a problem is found with the last piece inspection, the auditor is to inspect all circuits and determine which, if any, of the circuits are defective and must be scrapped or repaired.

Figure 10.6 Optimum Cutter Loading and Sequencing Report.

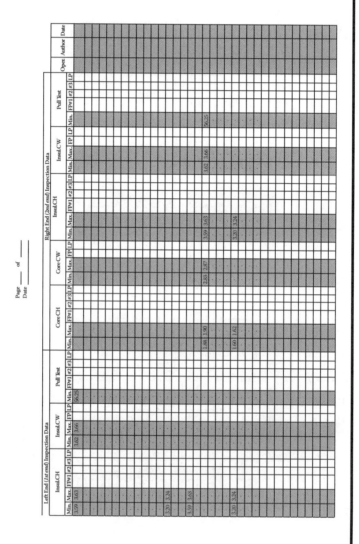

Figure 10.6 (Continued)

What Size Should the Cycle Quantity (Lot Size) Be? ■ 283

Figure 10.7 OCLASS Inspection Report.

this quantity, they merely try to adhere to the objective of minimizing cut quantities because this is what most of the "experts" recommend in recent books on Lean Manufacturing. The cost of inventory is only one of many factors, some of which have a much bigger impact than inventory, that determine the overall cost of cutting. Fortunately, it is not overly difficult to make a scientific determination of the Optimum Cut Quantity (OCQ) based on an analysis of the entire cutting system. The system detailed below is such a system, and it is highly recommended that this system be used to make a determination of the OCQ based on facts and not emotion or opinion.

This system estimates the yearly cost of the entire cutting operation at different cut quantity levels. To understand how this is done, we first need to understand the various cost elements that comprise the yearly cutting cost. This program considers these cost elements in two buckets. One cost bucket is comprised of elements that will decrease in cost as the cut quantity increases because the uptime and output of the cutters will increase as setup time is decreased (fewer setups means less downtime and more uptime resulting in more cut leads per cutter). These elements are all related to costs associated with cutting machines, i.e., cost of investment dollars for cutters and dies, depreciation for cutters and dies, tools for cutters and dies, spare parts for cutters and dies, maintenance for cutters and dies, floor space for cutters, cutting labor, indirect support, and scrap generated at setup.

The other basket of costs is comprised of elements that will increase in cost as the cut quantity increases. These elements all relate to cut lead inventory, i.e., the cost of inventory dollars, the cost of managing inventory, the cost of investment dollars for racks, and the floor space required to house inventory on the racks.

Note: The assumption is made that the amount of scrap material generated as a result of improper setups or defective material will not change as cut quantities increase. Fewer setups means fewer setup errors, but more circuits may be scrapped per occurrence. The assumption also is made that the warehouse inventory will not change due to changes in cut quantities.

This system adds these two buckets of costs together to obtain the total cutting cost (at various specified cut quantity levels). If we assign Yearly Cost as the y-axis and Cut Quantity as the x-axis, at a very low cut quantity (e.g., 200 pieces), the yearly cost of the entire system will be very high. This is due to the fact that a large number of cutters are required, because there is such a high percentage of available cutting time allocated to setup and associated delays.

As the cutting quantity is increased, the yearly cost of cutting will decrease rapidly at first (because, initially, we are increasing the percentage of raw cutting time rapidly because the raw cutting time base is so low). This will occur because the decreasing costs associated with the cutters will decrease in much larger dollar amounts than the increasing costs associated with inventory will increase.

As the cut quantity continues to increase, this decreasing cost trend will continue, although at a less abrupt rate, until the curve will start to flatten out. The cost curve will be relatively flat for a series of cut quantities that may span several thousand pieces. In this flattened area, the curve will dip a little as an additional cutter comes out due to some increased raw cutting time, and then start gradually increasing (due to the impact of increasing inventory) until the next cutter comes out. Eventually, it will not be possible to eliminate any additional cutters because of the very low percentage of additional raw cutting time made available (because the base of raw cutting time is now high) by further increasing cut quantity size. The curve will then gradually climb (without further dips) as the inventory continues to increase with no offsetting reduction in costs attributable to reduced cutters (Figure 10.1).

The Optimum Cut Quantity (OCQ) is based on number of pieces and not days of cutting. Volume has some impact on the OCQ (higher harness volume usually means a somewhat higher OCQ), but it is not the major driver. For the cost data that are currently input into the system, the OCQ will generally be in the 3500+/− range. For very high-volume packages, the OCQ might be 7000 to 10,000, but the difference in yearly cost between a cutting quantity of 7000 to 10,000 versus one of 3500 may not amount to a tremendous amount of money. If so, we can potentially justify the lower cut quantity, but, at least we will have the data to make sound decisions instead of just guessing.

Note: Regardless of the calculated OCQ, it is not a good idea to have cut lead inventories in excess of 15 days (or even less, depending on the customer), due to the problem of reacting to potential engineering changes and the risk for obsolete leads. Because the OCQ is based on the number of pieces and not the number of days of cut lead inventory, the number of days of cut lead inventory established by the OCQ may vary from a week or less for very high-volume packages to up to three weeks for low-volume packages.

The attached OCQ Report (Figures 10.2, 10.3, and 10.4) should be generated within a plant for a specific cutter type, as well as for cutters that will apply seals versus those that will not (and those that will apply doubles versus those that will not), but the report can be run for any subset of business.

The first block of datum or *Assumptions* on the attached report (Figure 10.2) is information that is determined for each type of cutter used in the operation. Studies and analysis of data will be required to establish this information. The information will need to be updated from time to time due to changing conditions and continuous improvement. The highlighted numbers are computer calculations made from the data that are input.

The lower block of datum (Figure 10.4) contains all of the pertinent calculations needed to determine the optimum cut quantity—calculations made from the data in the upper block. Those calculations highlighted in yellow are of a particular interest to a cutting operation. All of the calculations highlighted in green are yearly cost numbers that are added together to determine the yearly departmental cutting cost for the various cutting quantities indicated.

The graph depicts the yearly cutting costs for each of the cut quantity options. The cut quantity that gives us the lowest total yearly cost is 8000 leads (almost 23 days average inventory), but this generates only a $14,000 savings versus a cut quantity of 5000 (a little over a 14-day average inventory). For a yearly cost penalty of about $230,000, a cut quantity of 3000 could be utilized (an average inventory of about 8.5 days). Depending on the customer, either of the last two options is probably preferable to the first, but management has all the facts to make an enlightened decision. No longer must they make decisions on gut feel, opinion, or what someone has told them they should do. (Figure 10.3 shows the yearly department costs based on various cut quantities in the graph form.)

Optimum Cutter Loading and Sequencing Subsystem (OCLASS)

The Optimum Cutter Loading and Sequencing Report are designed to be the only documents that are required in the Cutting Department to provide all of the information needed to properly set up, process, and verify circuits. If this system is properly utilized, along with the Optimum Cut Quantity (OCQ) Subsystem, it will guarantee that the lowest overall cost will be achieved for cutting wires.

Within this subsystem, each circuit is loaded onto the optimum cutter based on gauge, length, and applications. The circuits are then sequenced to minimize the setup time. The required daily cutting hours are automatically calculated for each circuit, then each cutter is loaded to the number of hours established by policy.

The optimum sequencing and cutter loading contained within this subsystem is fairly simple and straightforward. Philosophically, what we want to accomplish is to load each circuit (lead) onto the cutter type on which it can be processed reliably at lowest labor standard, and once each circuit is loaded onto a specific cutter type, we want to sequence the circuits on each cutter such that the minimum setup time is achieved. In other words, once a circuit is listed on the report, we want the next circuit that is listed beneath it to be the one circuit requiring the least amount of setup time among all circuits in the entire cutting area.

How do we make this happen? We could assign an amount of time required to accomplish each potential aspect of a setup, i.e., seal applicator change, seal change, terminal die change, terminal change, strip change, wire gauge change, wire type change, wire color change, and length change. Once this was done, we could select a starting circuit and compare it to every other circuit in the plant, and find the circuit requiring the least amount of setup time. We could continue in this manner until all circuits had been placed in the sequence. However, it would be almost impossible to obtain completely accurate data and this would require a tremendous amount of computer time to accomplish. Fortunately, there is a much easier, cheaper, and less time-consuming way to accomplish the same thing.

We still need to determine the amount of time required to perform every aspect of a setup. Once this is done, we can prioritize the steps from most time consuming to least, remembering that when some setup steps are done, it implies that one or more other steps must be done. For example, it takes less time to change a terminal die than it does to change a terminal, but, when we change a terminal die, it also means that the terminal will be changed. Therefore, "terminal die change" will precede terminal change in the sequence. We also know that a given terminal die, and all terminals that can be terminated on that die, will always use the same strip length. We, therefore, can move "strip length change" ahead of "terminal die change" in the sequencing priority for single leads, which will ensure that we rarely, if ever, have to change a first-end strip length.

If circuits are sequenced with these priorities, remembering that there is both a first-end and a second-end potential for die changes, seal changes, terminal changes, and strip length changes, we can achieve the minimum setup time requirement.

Note: Doubles and seals can only be applied on certain cutters, thus, in order to route these circuits/assemblies to the proper cutters, doubles and seals are placed first on the priority list. Seals are more time consuming to set up than doubles, but doubles are prioritized first because some doubles contain a seal on one, or possibly both, of the other ends.

Understanding how the sequencing is accomplished, the process is started by the computer determining the type of cutter on which to load each circuit, based on the cutter types available within the area. This is easily accomplished by loading the capabilities of each type of cutting machine into the database. As the computer looks at each circuit, it will automatically route the circuit to the optimum available cutter (lowest overall labor standard given compatibility and quality). The computer also identifies the next best cutter type in case of an overload on the optimum cutter type.

Once each circuit is routed to a cutter type, the computer then loads each cutter (to the established number of hours) with circuits listed in the optimum circuit sequence (as explained above). This system is designed to prioritize the various setup steps by volume, higher volume being considered first. To maximize the advantage of volume, the computer considers the highest volume applications to be the first end, regardless of the blueprint orientation.

For example, for seal applications, the highest volume seal die will become the first end if there are seals on both ends of the circuit. If there is only one seal, it will automatically become the first end. For nonseal leads, if there is only one terminal applied on the cutter, it automatically becomes the first end. If there are two terminals or no terminals applied, the first end becomes the end with the highest volume strip length. Doubles are always considered the first end.

Once this is accomplished, the computer is ready to start loading specific circuits to specific cutters. All circuits with doubles are loaded onto the designated cutters first. The highest usage doubling die is prioritized first; this is followed by the double strip, then, in order:

The double terminal
Circuit A seal applicator
Circuit B seal applicator
Circuit A seal
Circuit B seal
Circuit A strip

Circuit A die
Circuit B strip
Circuit B die
Circuit A terminal
Circuit B terminal
Circuit A gauge
Circuit B gauge
Circuit A insulation type
Circuit B insulation type
Circuit A wire color
Circuit B wire color
Circuit A length
And finally, the circuit B length.

Circuit A and Circuit B are determined by volume, the same as the first and second ends are determined.

Note: Center strip circuits also must be directed to specific cutters, as with doubles and seals. The remaining sequencing priorities for circuits that are center stripped are the same as for other circuits.

Next, all circuits with seals will be loaded onto the designated cutters, starting with the circuit that has the highest volume first-end seal die, and then within this group, the highest volume second-end seal die, then first-end seal, then second-end seal, then first-end strip, then first-end die, then second-end strip, then second-end die, then first-end terminal, then second-end terminal, then gauge size, then wire type, then wire color, and, finally, length. With this sorting, there may be several circuits in a sequence that have nothing more than a simple length change (listed longest to shortest). After these types of circuits are exhausted, the computer will sequence circuits that are the same except that they require a color change (listed highest volume first) and probably also a length change. Once these circuits are exhausted, the computer will sequence circuits that are the same except that they require a different gauge (and probably a different color and a different length).

This process will continue until we have sequenced all of the circuits that utilize the highest volume first-end seal die. We then go to the circuits that require the second highest usage left end seal die and follow the same process.

For circuits that do not require seals, the same process is followed, except that we begin with the circuits that have the highest volume left end strip. The sequencing priority is: Left end strip, left end die, right end strip, right end die, left end terminal, right end terminal, gauge, insulation type, wire color, and finally length (longest to shortest). With this sequencing philosophy, it is probable that several cutters will never have to change the left end strip, and some will not change left end dies; possibly not even left end terminals.

The computer will follow this sequencing philosophy and start loading cutter #1 (for each cutter type) until it has reached the established loading level (in standard hours per day). Once this level is reached, the next circuit in the sequence is moved to cutter #2. This cutter is loaded utilizing the same philosophy until we reach the loading level on cutter #2, at which time the loading starts on cutter #3, etc.

If there is insufficient capacity on a certain type of cutter to load all of the designated circuits, the excess circuits will flow over onto the second best cutter type.

With the above philosophy, many of the cutters will be loaded with circuits with high-volume applications that will have minimal first-end changes. A few cutters will be loaded with lower volume applications that will require more extensive changeovers. One or more of the cutters will be only partially loaded, or completely unloaded, depending on the planned tooling capacity in the cutting area and the philosophy on the loading of individual cutters.

Note: There is some wisdom in loading most of the cutters to 100% (based on the standards and delays that are in the database, which have been developed through extensive studies) and having all of the excess capacity available on a few cutters in the area, which are to be run by team leaders or other specialized operators. The problem with having excess capacity on each machine is that operators will have a tendency to pace themselves, thereby losing the extra capacity. By loading the cutters to 100%, it is also much easier to identify problems with operator performance, standards, or delays. This enables the department team to resolve the problems much more expeditiously.

Ideally, we would put our best operators on the cutting machines that have excess capacity, which are loaded with the "cats and dogs." These highly qualified operators can help provide flexibility if problems arise on any of the other cutters, provide the flexibility to allow the temporary shift of circuits from cutters requiring preventive maintenance, provide flexibility to help train new operators as needed, and provide flexibility to accomplish other tasks in the department as assigned by the foreman.

Chapter 11

Wrap-Up: How about "Intelligent Manufacturing" for Real Change in Which We Can Believe

Hopefully, by now you should have a good idea of some of the things that will be highlighted in this wrap-up. I'm sure many of you with on-the-floor manufacturing plant experience feel as I do. We have been overwhelmed with Lean Manufacturing concepts that have been misunderstood and/or misapplied within our respective companies and have created anything but "lean" conditions.

My proposal is that companies implement the concepts of "Intelligent Manufacturing" that, in a nut shell, is learning from the mistakes and successes of others and doing things that just make good common sense and which can be justified by real savings and not by imagined improvements. Below is a summary list of Intelligent Manufacturing concepts that can unquestionably help any company, especially a manufacturing company, become more successful:

- **Quit doing dumb stuff.** Most of the really dumb things that were done by Packard, some of which I wrote about in this book, were inexcusable. They should have never been done, at least not on the scale they were. Many of these mistakes were made because someone read a book or listed to an "expert" who knew a lot less about what

Packard needed to do to be a successful manufacturer than virtually anyone working within the Packard operations. Others were forced on the company by someone trying to make a name for himself by implementing an "innovation" that would surely be noticed by upper management. "Change for change sake" can be a very dangerous proposition, especially when the ulterior motives are for personal gain and not necessarily for what is in the best interests of the company. Companies can help protect against this type of problem by putting a little less emphasis on innovation and a little more emphasis on excellence, especially obtaining the highest levels of productivity and quality possible with best current technology before exploring uncharted waters.

- **Be careful who you listen to.** This goes along with the prior point. My experience is that most consultants and so-called experts hang their hats on one or two concepts, which they attempt to "sell" to a company with a promise that these concepts will convert the company into an overnight success. Most so-called experts seem to be long on theory, but short on tangible successes (or any real experience in a manufacturing environment, for that matter). There are some good consultants and some potentially beneficial ideas and benefits out there, but if some advice doesn't seem to make sense, it probably doesn't. Peter Drucker said that "management is doing things right; leadership is doing the right things." Packard had way too many managers and not nearly enough leaders. Unfortunately, a lot of other companies do as well.

- **No company will have long-term success without good leadership**, so if you don't have it you had better encourage and develop it. Drucker said that effective leadership is not about making speeches or being liked; leadership is defined by results, not attributes (which is bad news for the "empty suits" of the business world, who are so rife in many companies, including Packard). Leaders are not easy to come by, especially due to the shabby way they are treated in many companies. Two of my favorite sayings fit well here: (1) don't confuse efforts with results, and (2) it's not how much you know, but how much you accomplish with what you know that is important.

- Now is probably a very good time to talk in a little more depth about leaders, at least the type of leaders who will make a positive impact on any company lucky enough to have them. I recently read an outstanding book by Jim Collins entitled *Good to Great* (Harper Business, 2001), and his extraordinarily well-researched work really helped clarify for me many things which I had observed throughout my working career

and believed to be true about truly successful leaders but didn't have the extensive data to confirm as such.
- Collins wanted to know if it was possible to turn a good company into a great company, and, if so, how it could be accomplished. He and his team embarked on a quest to find the most successful company transitions over the past half century. What he discovered were 11 companies which had averaged cumulative stock returns 6.9 times the general market in the 15 years following their transitions, performance that outshone all other S&P companies being evaluated by a large margin including Wal-Mart, Coke, GE, McDonalds, and many other household names.
- They were quite surprised at the things which, based on their research, failed to have a significant positive impact on the transition of these companies from good to great. There were no white knights coming from the outside to rescue these going-nowhere companies (in fact, there was a negative correlation when white knights were brought in). The form of executive compensation also seemed not to be a factor (another great reason to avoid bonus compensation based on short-term results). There was also no evidence that superior strategy, the introduction of sensational new technologies, or mergers and/or acquisitions played an advantageous role. These companies were not in great industries, in fact, many were in terrible ones based on the overall market conditions at the time. None of these companies did anything special to kick off their transition periods. There were no announcements with big fanfare, no wasteful team building seminars, no major realignments, and no big launches. They all just started on a steady climb to the pinnacle.
- So, what was the major factor in every one of these 11 transitions to greatness? It all boils down to one word—leadership. But not just any leadership. It is the kind of leadership that Collins and his team labeled Level 5 leadership. All of these companies were led by Level 5 leaders as they defined this type of leader, and the companies, as a result, put real efforts into identifying and developing more of them, and recognizing and rewarding them appropriately. If you want to visualize the characteristics of this type of leader, just imagine the polar opposite of what we currently have occupying the White House.
- This type of leader is modest to the extreme and shuns public adulation but has tremendous resolve and is fanatically driven to produce results. He motivates through inspired standards and not through personal

charisma, and he has the determination to overcome all obstacles to achieve the desired outcome. He channels all of his energy into the company and sets up his successor for even greater success. He readily accepts blame when things do not go right within the company, but he is extremely generous in appointing credit for successes to other people, external factors, and to good luck. (See, I told you how you could easily visualize this type of leader.)

- Collins writes, "Look for situations where extraordinary results exist but where no individual steps forth to claim excess credit. You will likely find a potential Level 5 leader at work."
- I worked with several individuals meeting the Level 5 leadership criteria in my tenure at Delphi Packard, but it was rare when one of them made it very far up the corporate ladder. Packard (and as best I could determine, GM) were much more impressed by style over substance, at least over the last several decades where I was able to observe (although, there must have been some very fine Level 5 leaders both within GM and Delphi in the early days to have created the outstanding companies they once were).
- What was more often the case, at least within Packard, was that the golden boys and girls were given abundant credit for successes over which they had little positive impact; but, when they were right in the middle of things which went horribly wrong and for which they were largely responsible, the blame landed elsewhere (sounds a lot like how things are going in our country right now, doesn't it?), so that their advancement potential was not compromised. This is the formula for how a true leadership void can be created and persists at the top of an organization, a void which will ultimately spell doom for any company.
- **If you want to control critical factors in a manufacturing environment (or anywhere else for that matter), you had better find a way to measure performance.** What is not measured is not controlled; just look around for verification. If you want to jump-start performance, report to the right people on the performance that is being measured. Just make sure the measurement tools are objective, fair, simple and straightforward, and capture what is really important (i.e., key performance indicators that will drive the bottom line).
- **For most manufacturing operations, direct labor is the most critical and the most costly factor and it must be controlled.** Not having a good system to control direct labor is a recipe for failure,

as is not having proper manufacturing department organization and supervision. This needs to be realized throughout the organization, and appropriate company resources need to be devoted to the control of direct labor to ensure that this most critical of all resources is utilized effectively.

- **You will get the performance you motivate (at least from the overwhelming majority of your employees).** If your policies and practices punish good behavior or reward bad behavior, the outcome is absolutely certain—the laws of nature will not be violated. If you want confirmation, just look at public education or virtually any government program. If you want more of something, then reward it; if you want less of something, then punish it. It is not any more complicated than that. Just make sure care and good judgment are utilized when using the carrot or the stick to make sure that the rewards or punishments are appropriate and will garner the results you seek without creating precedents or unintended consequences that will be hard to manage or that will create future headaches.
- **Proposed changes to current systems, processes, or procedures, as well as new projects and programs, must be justified through objective scientific analysis.** This means that you don't work backwards to get the answer you want, rather you allow vigorous and thorough scientific analysis to dictate the correct answer, even if it is not the answer you were hoping for. The computer can be a great tool to help this become a reality, as well as doing a thorough analysis in an engineering environment, such as a methods lab or a prototype/pilot area. Of course, it is always a good idea to use realistic assumptions, and if we err, it should be on the conservative side when we are trying to justify something new and on which we have no real working experience or history.
- **Think before you act and make sure that what you are proposing to do makes good common sense and can be justified with tangible benefits**. This point goes along with a couple already mentioned, but it is a good concept to keep in mind at all times. Drucker also said that taking action without thinking is the cause of every failure, and he was right. Even if the house is burning down, a few precious seconds of thought might save the day.
- **Do not install any new manufacturing process or system into the plant that has not been thoroughly vetted, debugged, and justified in a controlled pilot program**. If only this were done for all government programs, how much better off we would all be. In fact,

that is the system our Founding Fathers put in place with Federalism and the Tenth Amendment. Our federal government will probably continue to make the same mistakes, but companies do not have to. Once the new process or system is thoroughly proved and justified, it should become the new standard for the entire operations and proliferated wherever applicable.

- **There may be a lot of ways to accomplish a job, but choose one and implement it throughout the operations.** Standardization is critical to enhance control, reduce costs, enable good communications and training, enable the effective transferring of employees, etc. Often, there is one best way of doing something. If so, this way needs to be adopted throughout the operations as the standard. When there is more than one way of doing something which are equally good, choose one and standardize it throughout the operations.

- **The KISS (keep it simple, stupid) principle should be used when developing manufacturing systems.** This includes all processes, layouts, material handling and flow systems, and methods within the various departments. The more complicated a manufacturing system is, the less likely it is that it will be managed and run effectively. The more complicated a system is, the more opportunity for errors there are, and the more errors there will be.

- **All manufacturing processes and systems must be analyzed and designed through the use of scientific tools**, e.g., through the use of preplanning or other types of engineering studies and evaluations. The end result will be well defined processes and systems that are complete with detailed methods, accurate standards, quality responsibilities, layouts, and material flows into and out of the processes. Virtually every problem that can occur in production can be anticipated and simulated in a proper engineering evaluation and corrective or preventive steps can be put into place to avoid major manufacturing disruptions where problems become very expensive, not only in dollars and cents, but in customer relations, standing in the industry, and in the attitudes and motivations of employees.

- **Right size your lot size.** Lot sizes (cycle quantities) should be determined scientifically and not by trying to adhere to the latest fad. Because setups cost money, often a lot of money (often in the form of downtime, setup labor, maintenance, scrap, extra material handling, additional supervision, product identification and segregation, etc.), they should be justified. Unless respective materials are being delivered more than once a day, it will be rare that a cycle quantity of less than one day's requirements will be justified. Depending on the costs associated with setups, cycle quantities of several days, or even a week or more, may be the most economical. Run the numbers and let them be your guide.

- **Utilize paced operations, where appropriate.** Paced operations are not always appropriate due to such factors as product design or complexity, volume requirements, tooling requirements, etc. However, if a product does lend itself to a paced operation, in almost all cases, the manufacturing system should be so designed (thank Henry Ford for this decades-old concept). Paced operations provide for much better labor control for both productivity and quality, and they generally result in cost savings related to reduced direct labor content, reduced indirect labor and supervision, and reduced investment.
- **Avoid suboptimization like the plague.** Before any element within a system is changed, the impact on other interrelated elements within the system needs to be evaluated. Inventory and salary heads are two cost elements within a manufacturing system that are notorious for being suboptimized. Optimizing one of these elements, or some other element, within the system without considering the impact on the system as a whole could lead to a deprovement within the entire system, a loss of confidence in management, and a lot of heartburn for everyone involved.
- **Slow and steady progress is the key to success.** Companies that spend all their time trying to hit home runs will strike out most of the time and lose their focus on the basics of the business, very often backsliding in the process. Companies that focus on the basics, looking for modest but continuous improvements, tend to continually improve, and occasionally they even hit a home run. Also, think about the impact on the employees within a company, both hourly and salary, when a company is in constant turmoil because it is continually trying to reinvent itself.
- **Don't assume that the old way of doing things, which is now out of favor, is necessarily wrong**, nor assume that the new, popular way of doing things is necessarily right. Over the past several decades, many new concepts have come and then have gone as quickly as they came. The wheel is still with us. Don't follow the in-crowd; perform thorough analyses on any new concept you are tempted to employ to see if there is a payoff, which there must be if you are going to exercise the organization. And, just because some new concept apparently

works for someone else, whose circumstances may be entirely different than yours, it doesn't mean that it will work equally well for you.

- **Make computers work for you, not the other way around.** Some generic programs can be helpful to do some things for some companies, but, the only way that most manufacturing companies are going to obtain computer programs or systems that help drive improvements and gain a competitive advantage is to design their own programs specific to their needs. Depending on circumstances, the programming can be contracted out, but my experience is that it is usually best to do your own programming and program maintenance, if possible.
- **Be very careful what and who you benchmark.** Benchmarking can be helpful, but remember that concept you are benchmarking may not be as good as what you are currently doing, at least not for your specific needs. Just because a company might be successful (by all outward appearances), it does not mean they have it all figured out and that all of their processes and systems are worth emulating. And, even if something works well for someone else, it does not mean it will necessarily work equally well in your company, which, in all probability, has different products, resources, policies, procedures, and culture.
- **Never add nonvalue-added work to your company, at least, not unless there is some overriding principle involved.** If an activity does not reduce cost, improve delivery, improve quality, add to the knowledge base, or significantly improve the work environment, it should not be done.
- **Use a tool as a means to an end, not as an end in itself.** There are some very good tools out there that can help a company resolve specific problems or improve performance in various areas (several are presented in this book), but remember that these are tools to help accomplish an objective. Making the objective the implementation of the tool is putting the focus in the wrong place and the results will undoubtedly be very disappointing.
- **When a company is contemplating spending any significant sum of money, key evaluators and decision makers should use the following decision-making criteria**: Pretend that you personally will receive the benefits of the contemplated change, but also pretend that it is your money that will be used to implement it. If you would not be willing to spend your own money to implement a project, you should not be willing to spend the company's money either. Too bad our politicians seldom use this decision-making criteria.

I could go on, but I believe this wrap-up has captured most of the important concepts which have been highlighted in this book. My goal in writing this book is to help other manufacturing companies, especially U.S. companies, become more competitive by learning from the failures and successes of GM and Packard, especially since these companies did at one time have a very good handle on things, and by learning from the experiences of someone who has had a great deal of on-the-floor operations experience, both nationally and internationally. I hope I have succeeded in this endeavor.

Index

A

absenteeism control, 226–231
Ackerman, Dan, 12
advice, manufacturing supervision, 158–159
Agents of Influence, x
agricultural practices, xi
Alaska National Wildlife Refuge (ANWR), xvi
alternative energy sources, xvi–xvii
American Idol, 55
assistant foremen, 158
automated productivity system, 188–189
automation, *see also* Technology
   failure of, 59–66
   overview, 51–53
Automotive Components Group, 36
Auto World, 26

B

base standard, 190
base standard adjustment (BSA), 190
batch processing, 121, 123
BECs, *see* Bussed Electrical Components (BECs)
behavior, *see* Performance
benchmarking, 102–105
"best manual method," new systems comparison, 67
better *vs.* consistency, 16–17
Black Friday, 33–36
blame landing elsewhere, 294
Boeing, 79–80
Boston Celtics, 26
Bowen, Kent, 16, 85–86, 98, 123, 199
BSA, *see* Base standard adjustment (BSA)
Buchanan, Pat, *x–xi, xii*
Budget DLB, 185–188
budget routing, 186–188
build sequence, 211
Bussed Electrical Components (BECs), 121

C

capabilities, lack of difference, *xviii*
capitalism system, *xv–xvi*
career path placement, 28
carrot and stick method, 295
challenges, cycle quantities, 267–270
changes
   affecting leads, 247
   to be implemented quickly, 247
   for change's sake, 297
   common sense, 295
   coordinated, 246
   engineering change implementation, 240–241
   payoff analysis, 297–298
   requiring new materials, 246–247
   scientific analysis justification, 295
   tangible benefits, 295

Chapter 11 protection, 40–43
cheap labor *vs.* productivity, 151–152
Choate, Pat, *x*
Churchill, Winston, 62
circuit test sequence, 211
Collins, Jim, 292–294
common sense, 295
compensation packages
   current state, 13–14
   termination, 41–42
   tied to self-interest motivation, 8–9
competitiveness, *xviii–xx*
competitive position, *xii–xiii*
Components Methods Lab, 80–81, 150
component subassembly, 195
computer systems, 219–235
concepts, transfer of, 143–147
Concepts of the Corporation, 10
consistency *vs.* better or smarter, 16–17
consumers, corporate taxes, *xiv–xv*
content of work, 98, 109
continuous flow, creating, 87–88
continuous improvement
   learning organization, 96–97
   standardization, 91–92
controlled pilot program, 295–296
controlling productivity, 184–185
control sheets, 208–209
conveyors, 59–70, 173
corporate taxes, *xiv–xv*
costing/pricing system, 40
country of origin, *xviii–xx*
Cowell, Simon, 55
credit, excessive, 294
critical factors
   controlling through performance measurements, 294
   direct labor, controlling cost, 294–295
critical four M relationship, 207–208
critical value streams, 239–240
customer-supplier connections, 98–99
cycle quantities
   challenges, 267–270
   implementation, 263–267, 270–272
   optimum cut quantity, 272–280, 284–286
   optimum cutter loading and sequencing subsystem, 286–290

overview, 259–263, 279–290
right sizing, 296–297

D

Da, *see* Department allowance (DA)
"Death of Manufacturing," *x–xi*
decisions
   background, 7–11
   consensus, 96
   long-term philosophy, 87
   reasons for doing, 291–292
   smart people, 49–82
"Decoding the DNA of the Toyota Production System," 85, 199
DeLorean, John, 10, 64
Delphi Automotive Systems, 36
Deming, Edwards, 11, 24, 114, 183
department allowance (DA), 191
department layout, 215
dependency on foreign suppliers, *x–xi*
Dephi Automotive Corporation, *xx*
Dephi Corporation, *xx*
Dephi Packard, *xx*
design analysis, 209–210
design for manufacturability review, 210
direct labor, controlling cost
   critical factors, 294–295
   manufacturing support, 150
Direct Labor Bibles
   example, 192–197
   labor estimates, 185–186
   overview, 162–164
direct operator time transfer sheet, 175
downtime sheet, 176
Drucker, Peter, 5, 10–11, 79
dumb decisions, *see* Decisions
duplicate boards, 216

E

Einstein, Albert, *xii*
"empty suit," 10
experiences, learning, *xx*
experts
   careful selection, 291–292
   no substitution for thinking, 120

F

fair trade and fair trade practices, *xiii*
Fanny Mae, 9
figures and liars, 53
final breakdown meeting, 215
Flint, Jerry, 7–8, 9, 11
floor presence, 95–96
flow diagram, 215
Forbes Magazine, 7
Ford, Henry, 5, 23–24
foreman, 156
Forward Lamp Harness, 54–55, 59–70
four M relationship, 207–208
free market, *xxi*
free trade and free trade practices, *xii, xiii*

G

Gemba walks, 95–96
General Motors
    current state, 12–15
    historical developments, 4
Gilbreth, Frank Bunker, 23–27
gimmicks, 125
"golden" employees, 294
Goldratt, Eliyahu, 75–76
Good to Great, 292–294
Green Bay Packers, 19, 103
"green" energy sources, *xvi–xvii*
group leaders, 158

H

Hamilton, Alexander, *ix–x*
Harvard Business Review, 16, 85, 199
Henderson, Fritz, 12
HIPS, *see* Hybrid Integrated Production System (HIPS)
hiring quotas, 33–36
historical developments
    manufacturing, *xi*
    Packard Electric, 3–7
hourly employee classification matrix, 160
Howard, Tim, 9
Hughes Electronics, 79–80
human nature, 139–143

human self-interest, 8–9
Hybrid Integrated Production System (HIPS), 66

I

IDC, *see* Insulation Displacement Component (IDC)
ideal *vs.* realistic system, 123–124
implementation
    cycle quantities, 263–267, 270–272
    new concepts, 297–298
Industrial Engineering program, 43–47
industry practice, preplanning, 200–202
installation timing, 295–296
Insulation Displacement Component (IDC), 77–78
Integrated Production System (IPS), 59–67
international strategies, 37–38
International Union of Electrical Workers, 5
interview process, 2–3
inventory issues, 115–118
Inventory Reduction Project (IRP), 117
"Is GM Better or Ford Worse?", 16

J

Japanese vehicles, 6–7
job responsibilities, 156–157
just-in-time manufacturing
    inspiration for, 86
    overview, 118–120

K

key activity control, 242–248
kitting, 105–109

L

labor
    cost reduction, 28–29
    direct Labor Bibles, 185–186
    estimate, wiring harnesses, 209
    free market valuation, *xxi*
    performance and performance measurements, 186–188

standards, 216
vs. productivity, 151–152
leadership
  Black Friday, 33–36
  characteristics of, 93, 293–294
  long-term success, 292–293
Lead Prep Startup, 56–59
Lean
  benchmarking, 102–105
  core principles vs. implementation, 87–97
  inventory issues, 115–118
  just-in-time manufacturing, 86, 118–120
  kitting, 105–109
  one-piece flow, 122–128
  overview, 83–85
  Packard Production System, 100–115
  QS9000 implementation, 112–113
  Quality Circles, 110–112
  rules, 98–100
  summary, 128–129
  Team Leader Concept, 109–110
  Toyota Production System, 85–98
  U-cells, 120–121
  vista concept, 100–102
  zero defects, 114–115
learning from experience, xx
learning organization, 96–97
legacy, xxii
leveling the workload, 88–89
levels of operators, 148–149, 153–155, 160
Lewis, William, xix
liars and figures, 53
Limbaugh, Rush, xviii, xxi–xxii
line layout, 215
lines, justification for, 52–53
line start and follow-up, 216–217
Lombardi, Vince, 19, 103
Lordstown Assembly Plant, 73
lot size, right sizing, 296–297
low volume/part number proliferation allowances, 192

M

management, 27–29
manufacturing
  competitiveness, xviii–xx
  historical developments, xi
  scientific tools analysis, 296
  simplicity in development, 296
  system release, 216
manufacturing supervision
  advice, 158–159
  assistant foremen, 158
  foreman, 156
  group leaders, 158
  job responsibilities, 156–157
  levels of operators, 148–149, 153–155, 160
  overview, 150
  productivity vs. cheap labor, 151–152
  qualifications, 156
  support, 150–152
  time breakdown, 157–158
  wiring harness department, 152–160
master board release, 216
material availability, 231–235
McElroy, John, 16, 26
McKinsey Global Initiative, xix
measuring performance, 182–183, see also Performance and performance measurements
metal cutting, 193–194
methods lab, see Components Methods Lab
mistakes, 291–292, see also Decisions
model change notebook, 216
motivation, 7–11
Multicell systems, 121
multiple efficiencies, 168–169

N

national preferences, xv–xvi
new concepts implementation, 297–298
New United Motor Manufacturing, Inc. (NUMMI), 83, 109
nonpaced jobs, 145
nonvalue-added work, 102

O

Obama administration, 12, 293
Office of Federal Housing Enterprise Oversight (OFMEO), 9

OFMEO, *see* Office of Federal Housing
    Enterprise Oversight (OFMEO)
Ohno, Taiichi, 85
oil embargo (1974), 6
On a Clear Day You Can See GM, 10, 64
one-piece flow, 122–128
operator classification, 142–143
operator daily production report, 174
operators, levels of, 148–149, 153–155, 160
optimum cut quantity, 272–280, 284–286
optimum cutter loading and sequencing
    subsystem (OCLASS), 286–290
order repair tools, 210
outcome of work, 98, 109
overtime control, 226–231
overview, 4–6, 10

P

paced jobs, 145
paced operations utilization, 297
Packard Automobile Company, 3–4
Packard Electric
    background, 1–2
    historical developments, 37
    interview process, 2–3
    management, 27–29
    single customer, lack of competition, 1–21
    training program, 21–27
Packard Hughes Interconnect, 79, 220, 248
Packard Production System (PPS)
    benchmarking, 102–105
    kitting, 105–109
    overview, 100
    QS9000 implementation, 112–113
    Quality Circles, 110–112
    Team Leader Concept, 109–110
    vista concept, 100–102
    zero defects, 114–115
paper breakdown, 214
pathways for products, 99
pegboard review meeting, 210
performance and performance
        measurements
    automated productivity system, 188–189
    budget routing, 186–188
    carrot and stick method, 295

component subassembly, 195
controlling productivity, 184–185
conveyor daily production report, 173
critical factors control through, 294
Direct Labor Bibles, 162–164, 185–186,
    192–197
direct operator time transfer sheet, 175
downtime sheet, 176
example, 171–177
labor estimates, 186–188
measuring performance, 182–183
metal cutting, 193–194
motivation, *xiv,* 8
multiple efficiencies, 168–169
operator daily production report, 174
overview, 161–162
process and operator efficiencies,
    166–169, 179–180
production efficiency, 164–166, 178,
    184–197
productivity case study, 182
results, 169–170, 177, 181–182
rewarding the right behavior, 143–149
terms and definitions, 190–192
widget final assembly, 195–196
PI, *see* Plant improvement (PI)
Piggly Wiggly supermarket, 86
Planning Departments, 24–25, *see also*
    Preplanning
plant improvement (PI), 191
plant installation, 216
plant presence, 95–96
plant priority meetings, 132–133
plant standard, 190
power-and-free conveyor, 67–70
PPS, *see* Packard Production System (PPS)
prejudices, *xvi*
preliminary breakdown meeting, 214
preplanning, *see also* Planning Departments
    build sequence, 211
    circuit test sequence, 211
    control sheets, 208–209
    critical four M relationship, 207–208
    department layout, 215
    design analysis, 209–210
    design for manufacturability review, 210
    duplicate boards, 216

306 ■ Index

final breakdown meeting, 215
flow diagram, 215
industry practice, 200–202
labor estimate, 209
labor standards, 216
line layout, 215
line start and follow-up, 216–217
manufacturing system release, 216
master board release, 216
model change notebook, 216
order repair tools, 210
overview, 199–200, 208
paper breakdown, 214
pegboard review meeting, 210
plant installation, 216
preliminary breakdown meeting, 214
preplanning schedule, 209
rack order, 215
results, 205–207
schedule, 209
value of, 202–205
wire length letter, 216
wiring harnesses, 208–217
work content, 211–214
problems
   overview, 40–43
   proper manufacturing organization, 147–149
   quality culture, building, 89–90
   visual controls, 92
process and operator efficiencies, 166–169, 179–180
Production DLB, 186, 188–189
production efficiency
   nose dive, 184–197
   overview, 164–166
   report, 178
productivity case study, 182
Productivity Control System, 74–77
productivity vs. cheap labor, 151–152
progress, success factors, 297
proliferation allowances, 192
proper manufacturing organization
   advice, 158–159
   assistant foremen, 158
   foreman, 156
   group leaders, 158
   hourly employee classification matrix, 160
   job responsibilities, 156–157
   levels of operators, 148–149, 153–155, 160
   manufacturing supervision, 150–160
   overview, 131–132
   plant priority meetings, 132–133
   productivity vs. cheap labor, 151–152
   qualifications, 156
   reporting lines, changing, 133, 135–139
   solution, 147–149
   support, 150–152
   time breakdown, 157–158
   transfer of concepts, 143–147
   understanding human nature, 139–143
   wiring harness department, 152–160
proposed changes
   common sense and tangible benefits, 295
   payoff analysis, 297–298
   scientific analysis justification, 295
pull systems, 88

Q

QS9000 implementation, 112–113
qualifications, manufacturing supervision, 156
Quality Circles, 110–112
quality culture, 89–90
quotas, hiring, 33–36

R

rack order, 215
Raines, Franklin, 9
Rear Body Harnesses, 69
relentless reflection, 96–97
relief operator classification, 142
repair operator classification, 142
repair/scrap standard, 190–191
reporting lines, changing, 133, 135–139
right sizing, cycle quantities, 296–297

S

Saturn, 11, 84
schedule, preplanning, 209
scientific analysis and methods
   improvements mad, 99

proposed changes, 295
tools, 296
sequence of work, 98, 109
service conveyor, 61–66
service operator classification, 142
service standard, 190
silver bullet, *xx*
simplicity, manufacturing systems, 296
single-piece flow
    overview, 122–128
smarter *vs.* consistency, 16–17
Spear, Steven, 16, 85–86, 98, 123, 199
Spencer, Leanne, 9
standardization
    continuous improvement foundation, 91–92
    enterprise-wide adoption, 296
    manufacturing operations, 98
    parts, 5
style over substance, 294
suboptimization, 116–117, 297
successes
    leadership, 292–293
    slow and steady progress, 297
    total process cycle time, 255–258
    trying something tougher, 53–55
Superbowl winners, 26
suppliers
    necessity of, 30–33
    respecting, challenging, helping, 94–95
support, manufacturing supervision, 150–152
Synchronous Manufacturing, 72

T

tangible benefits, 295
tariff system, *x*
taxes, *xiv–xv*
Taylor, Frederick Winslow, 23–26
Team Leader Concept
    characteristics, 93–94
    overview, 109–110
technology, 93, *see also* Automation
terms and definitions, 190–192
The American Conservative, *x*
The Goal, 75–76
"The Platinum Rule," *xiii–xiv*

The Toyota Way, 87
Time and Motion Study, 26
time breakdown, 157–158
timing of work, 98, 109
total process cycle time (TPCT)
    benefits, 248–255
    changes affecting leads, 247
    changes requiring new materials, 246–247
    changes to be implemented quickly, 247
    coordinated changes, 246
    critical value streams, 239–240
    engineering change implementation, 240–241
    key activity control, 242–248
    overview, 237–239
    success, 255–258
Toyoda, Sakichi, 85
Toyota Automatic Loom Works, Ltd., 85
Toyota Production System (TPS)
    core principles *vs.* Packard implementation, 87–97
    GM incorporation, 15–17
    overview, 85–87
    rules, 98–100, 109
    Saturn outcome, 11
Trade Factor development, *xiv*
trade imbalances, deficits, and surplus, *xi*
training program
    Industrial Engineering program, 43–47
    origination, 23–27
    overview, 21–22
transferability, 119–120
transfer of concepts, 143–147
turnover control, 226–231

U

U-cells, 120–121
UCLA Bruins, 26
underutilization, 123
unfair trade practices, *xii–xiii, xiii–xiv*
unions
    caving into, 9–10
    current state, 12–15
    overview, *xxi*
    at Packard, 5

reason for management's decisions, 72–73
retirement packages, 42–43
universal operator classification, 142–143
U.S. energy policy, xvi–xvii
U.S. manufactured goods
   desirability, quality, cost, xvii–xviii
   manufacturing competitiveness, xviii–xx
utilization, paced operations, 297

V

value, preplanning, 202–205
value streams, 239–240
vista concept, 100–102
Visteon, 5
visual controls, 86, 92

W

Wagoner, Rick, 12
Ward's Auto World, 16
waste
   elimination, 129
   seven forms, 122–123
Whitacre, Ed, Jr., 12
white knight correlation, 293
Whitney, Eli, 5
widget final assembly, 195–196
wire length letter, 216
"wires bend," 5
wiring harnesses
   build sequence, 211
   circuit test sequence, 211
   control sheets, 208–209
   department layout, 215
   design analysis, 209–210
   design for manufacturability review, 210
   duplicate boards, 216
   final breakdown meeting, 215
   flow diagram, 215
   labor estimate, 209
   labor standards, 216
   line layout, 215
   line start and follow-up, 216–217
   manufacturing supervision, 152–160
   manufacturing system release, 216
   master board release, 216
   model change notebook, 216
   order repair tools, 210
   overview, 5, 208
   paper breakdown, 214
   pegboard review meeting, 210
   plant installation, 216
   preliminary breakdown meeting, 214
   preplanning schedule, 209
   rack order, 215
   specifications, 119–120
   wire length letter, 216
   work content, 211–214
work content, 211–214
workload, leveling, 88–89
World Wars, x
Wright, J. Patrick, 10

Z

zero defects, 114–115

About the Author

Even though I did a lot of writing during my 30-plus years as an employee of Packard Electric (now known as Delphi Packard Electric, part of the Delphi Automotive group that spun off from General Motors [GM] in 1999), especially for an engineer, I had never thought seriously of trying to write a book until fairly recently. This great country is changing rapidly and, unfortunately, not for the better in most cases.

I have a loving wife, five very responsible children, and 20 very special grandchildren, and my greatest fear is that they (especially my grandchildren and possibly even my children) will not have the same opportunities in coming years in this country as people my age have enjoyed. We are watching the free market system, which was established by our founders, being dismantled right before our eyes.

We are seeing irresponsibility rewarded while responsibility is punished. We are seeing multitudes of people, some of whom are in very responsible positions of power, claim that the Constitution is no longer a valid document, that it must be updated for the times. We are seeing countless demonstrations by people who have been blessed like no other people on the face of the Earth, but apparently have a disdain for this country and all for which it stands, or at least used to stand for.

I have very strong political views and I could go on and on regarding my concerns about the issues facing this country, but I realize that there are many great Americans who share my views and are in a better position than I am to sway opinion and affect public policy. However, I do have a background that does, I believe, enable me to provide some benefit to the country, especially in these very perilous economic times.

In an attempt to establish my credentials, I will start by saying that I grew up in a sleepy little East Texas town, spending my summers in the watermelon and hay fields earning a buck an hour. I did well in school, especially in math and the sciences, and excelled in sports. Upon graduation, I was offered football scholarships by several major universities. I ultimately decided to attend SMU in Dallas, because it was close to home and, more importantly, it had just started an Industrial Engineering program, which was of great interest to me.

Once on campus, I became overwhelmed and almost made the decision to go into business school instead. It seems that football players had not faired too well in engineering school over the years (well, to be fair, my dad was quite persuasive in helping me make the decision to stay in engineering). It turned out to be a good decision. I did pretty well on the ball field and I did quite well in school. I ended up making the Dean's list regularly and graduated with honors with a Bachelor's of Science in industrial engineering.

After interviewing with a number of companies after graduation. I decided to cast my lot with Packard Electric, which was a division of General Motors, located in Warren, Ohio. I was quite impressed with its massive operation. It was part of a very successful company (GM), it had a very strong Industrial Engineering Department, and it was located only about one hour from my parents' new family home, following my father's recent transfer.

I will be talking a lot more about Packard and GM later in the book, but for now let me state that Packard, in 1971, was a very strong manufacturing and engineering organization. I was fortunate to start in the Industrial Engineering Department as an industrial engineer and preplanner, and then spend the next several years in manufacturing and engineering supervision.

Industrial engineering was the center of the wheel at Packard, with spokes touching virtually all of the other functional areas within the company, i.e., manufacturing, application engineering, process engineering, plant engineering, quality assurance, quality control, material control, safety, maintenance, construction, and even finance and human resources management (HRM). Packard's application (or product) engineers designed all of the wiring harnesses for all GM vehicles, and industrial engineering was responsible for designing, developing, installing, and debugging the manufacturing systems for each of these products, in coordination with all of the other functional areas.

The Manufacturing Department is where one witnesses the culmination of all efforts in any manufacturing operation, and it is without question the most important of all departments within a manufacturing company. A manufacturing company can get by with some weaknesses in various functional areas, but if it has significant weaknesses in manufacturing, it will not be successful long term. I often related to others at Packard that I learned more about the wiring harness business and about relating to and managing people in the year and a half that I spent as a first-line manufacturing supervisor (foreman) than I learned in any other five-year period of my career.

After seven years in Warren, Ohio, I was named as one of the members of the Team of Ten, to be transferred to El Paso to start up a new plant in Juarez, Mexico (in 1978). We were initially told the idea was to start one plant of about 1000 employees, train Mexican counterparts, and then return to Warren. That was about 26 plants, 50,000 Mexican employees, and several hundred U.S. employees ago.

For a period of six years, I had industrial and methods engineering responsibility for three different plants as well as for the Mexican Supplier Program, to include developing and implementing training programs for industrial and methods engineers, developing and implementing a training program for production operators, developing forward plans, designing and installing a Methods Lab, starting up new business, leading efforts to identify and resolve plant problems, and a host of other functions.

I was next given the opportunity to manage all engineering activities in our newly acquired facilities in Portugal, which became part of Packard's European Operations. After two and a half very exciting and productive years in Portugal, I was asked to be the joint representative director for a new joint venture to be created in South Korea. This was a tremendous experience. We were able to start up a new facility of 500 people, with very little outside support, in a manner that exceeded Packard's expectations on timeliness, productivity, quality, and profitability.

After four years in South Korea, we returned to the United States in 1990. Over the next decade, I had a variety of interesting management assignments in Laredo, Texas; Foley, Alabama; and then back in El Paso, Texas. However, like our country, Packard had been changing dramatically over the prior three decades, and not for the better in most cases.

In 2001, after spending over 30 years with Packard, I finally grasped the reality that Packard was no longer the fine manufacturing company I had known in the early years. In fact, it was not even close, and likely would never be again because most of the people who had helped make it an outstanding manufacturer were no longer around. This was a bitter pill to swallow, knowing that Packard was at one time a shining star in what was to become the Delphi Automotive group.

During 2001, things began to look very bleak for Delphi Packard as well as for the entire Delphi Automotive Corporation. In an effort to reduce burden, Delphi offered significant incentives to highly paid salaried employees who would volunteer for early retirements. Due to my lack of political correctness (political correctness being an important factor in getting ahead at Packard), I knew I was never going to reach the highest levels of the company, so my ability to impact dramatic change on this ailing company was limited. I decided to take the opportunity to get out while the getting was good. Since then, things have gone steadily downhill for Delphi, and they filed for Chapter 11 bankruptcy in October of 2005. Delphi Corporation finally emerged from Chapter 11 the second half of 2009, almost four years after entering bankruptcy, but unless Delphi dramatically changes in the way they do business (which is highly unlikely), I wouldn't invest too heavily in this company if I were you.

After leaving Delphi Packard, I consulted for another large manufacturing company for several years before deciding to go into semiretirement a few years ago. The last few years have been a lot of fun, but I feel that there are things that I need to share while they are still fresh in my mind, for the benefit of any company that will objectively consider what is being offered.

My purpose in writing this book is to help other companies learn from Packard's and GM's mistakes—not only the mistakes of commission, but also the mistakes of omission. It is sad that some of the great concepts and systems that Packard had in place at one time were discarded instead of using renewed focus, and, in some cases, new technology, to enhance them. And, it is a shame that Packard lost the ability to differentiate between fluff and substance, even when it was offered to them on a silver platter. Hopefully, some of you reading this book will be able to do just that.

I also will be sharing several valuable concepts and tools that I developed during my working career of over three decades that I believe can help any conscientious company improve its competitive position and become a better company.